V International Forum for Young Scientists "Space Engineering"

AA003545

MATEC Web of Conferences Volume 102 (2017)

Tomsk, Russia
18 - 20 April 2017

Editors:

V. Borikov
S. Uchaikin
P. Baranov
V. Ivanova
A. Dolgih
A. Ignatovskaya

ISBN: 978-1-5108-3785-0

Printed from e-media with permission by:

Curran Associates, Inc.
57 Morehouse Lane
Red Hook, NY 12571

Some format issues inherent in the e-media version may also appear in this print version.

This work is licensed under a Creative Commons Attribution license:
http://creativecommons.org/licenses/by/2.0/
You are free to:
Share – copy and redistribute the material in any medium or format.
Adapt – remix, transform, and build upon the material for any purpose, even commercial.
The licensor cannot revoke these freedoms as long as you follow the license terms.
Under the following terms:
You must give appropriate credit, provide a link to the license, and indicate if changes were made. You may do so in any reasonable manner, but not in any way that suggests the licensor endorses you or your use. The copyright is retained by the corresponding authors.

Printed by Curran Associates, Inc. (2017)

For additional information, please contact EDP Sciences – Web of Conferences at the address below.

EDP Sciences – Web of Conferences
17, Avenue du Hoggar
Parc d'Activité de Courtabœuf
BP 112
F-91944 Les Ulis Cedex A
France

Phone: +33 (0) 1 69 18 75 75
Fax: +33 (0) 1 69 28 84 91

contact-edps@webofconferences.org

Additional copies of this publication are available from:

Curran Associates, Inc.
57 Morehouse Lane
Red Hook, NY 12571 USA
Phone: 845-758-0400
Fax: 845-758-2633
Email: curran@proceedings.com
Web: www.proceedings.com

TABLE OF CONTENTS

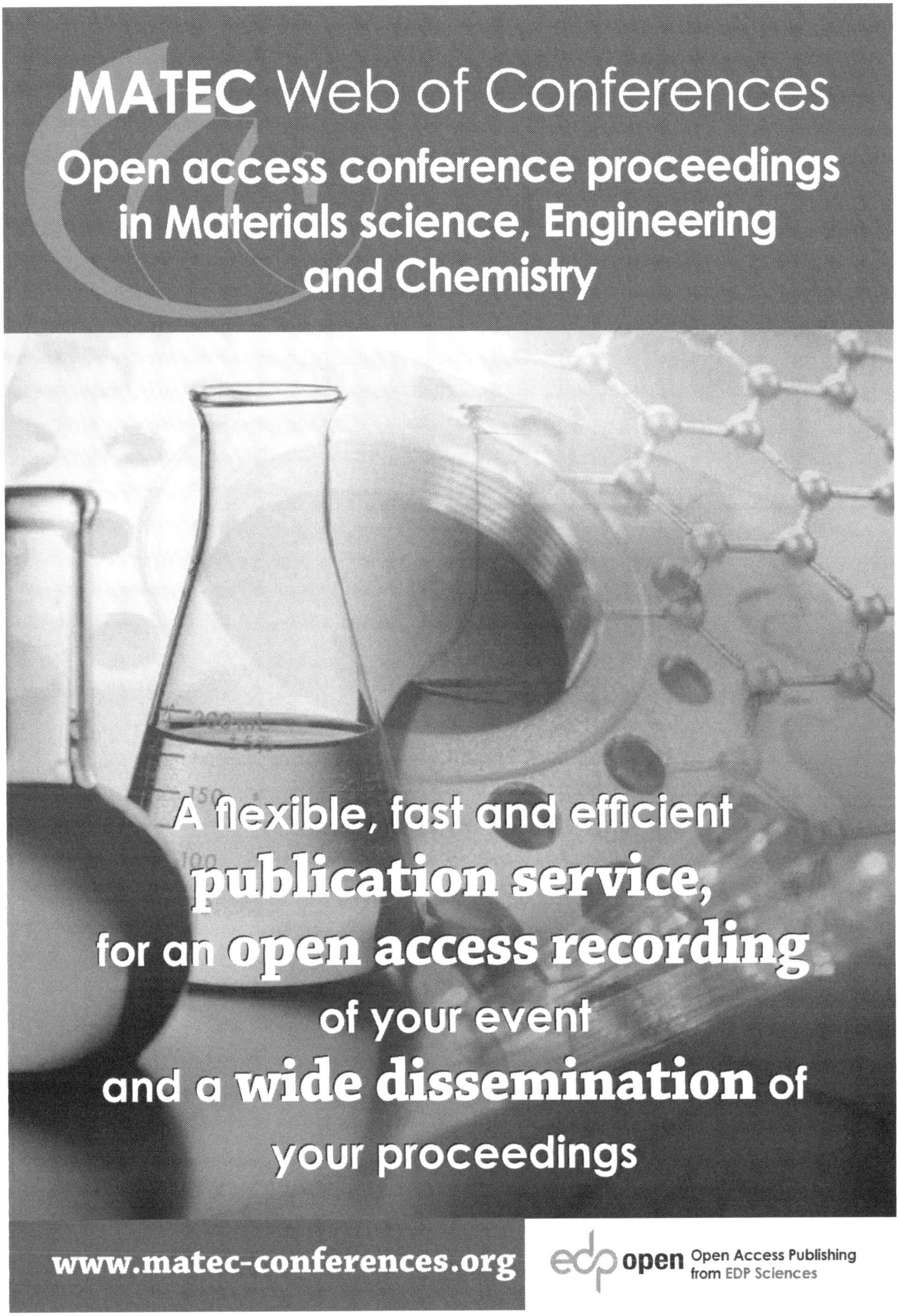
MATEC Web of Conferences
Open access conference proceedings
in Materials science, Engineering
and Chemistry
A flexible, fast and efficient
publication service,
for an open access recording
of your event
and a wide dissemination of
your proceedings
www.matec-conferences.org
edp open Open Access Publishing
from EDP Sciences

Aims & scope

MATEC Web of Conferences is an open-access journal devoted to the publishing of conference proceedings dealing with Materials science, Engineering and Chemistry.

The journal aims to make conference proceedings visible, citable and widely usable by the scientific community. It offers outstanding services for the publication and dissemination of contents in order to ensure long-term archiving and maximum exposure of scientific events.

Coverage includes all engineering disciplines: civil, naval, mechanical, chemical, and electrical engineering as well as nanotechnology and metrology. The journal also deals with the physical-chemical characterization of materials, their implementation, their resistance in their environment, etc. Others subdisciplines of chemistry, such as analytical chemistry, petrochemistry, organic chemistry, etc., and even pharmacology, are also welcome.

Why publish with MATEC Web of Conferences?

- **Open access:** conference papers available worldwide.
- **Fast publication:** less than 3 months.
- **Total flexibility:** no page restrictions, free unlimited use of colour in the online version, any kind of scientific document welcomed (traditional articles, posters, abstracts and slideshows).
- **High visibility:** all documents indexed in main databases and easily found online.
- **Fair author rights:** CC-BY Creative Commons license.
- **Range of additional publishing options:** printed proceedings, CD-ROMs, copy-editing, etc.

Contact: Isaline Augusto
matec@webofconferences.org

ISSN (electronic edition): 2261-236X

www.matec-conferences.org

edp open Open Access Publishing from EDP Sciences

Statement of Peer review

In submitting conference proceedings to MATEC Web of Conferences, I certify to the Publisher that I adhere to the **Policy on Publishing Integrity** of the journal in order to safeguard good scientific practice in publishing.

1. All articles have been subjected to peer review administered by the proceedings editors.
2. Reviews have been conducted by expert referees, who have been requested to provide unbiased and constructive comments aimed, whenever possible, at improving the work.
3. Proceedings editors have taken all reasonable steps to ensure the quality of the materials they publish and their decision to accept or reject a paper for publication has been based only on the merits of the work and the relevance to the journal.

Title, date and place of the conference

V International Forum for Young Scientists „Space Engineering"
18-20 April 2017, Tomsk, Russia

Proceedings editor(s):

Valeriy N Borikov
Sergey Uchaikin
Pavel Baranov
Veronica Ivanova
Antonina Dolgih
Anastasiya Ignatovskaya

Date and editor's signature 14.07.2017

17, avenue du Hoggar - PA de Courtabœuf – BP 112 - 91944 Les Ulis Cedex A (France)
Tél. : 33 (0)1 69 18 75 75 - Fax : 33(0)1 69 07 45 17 – www.edpsciences.org

Surface inspection problems in thermoelectric testing

Ahmed Abouellail[1,*], *Igor* Obach[2], *Andrey* Soldatov[1], and *Alexey* Soldatov[1]

[1]Tomsk Polytechnic University, 634050 Tomsk, Russia
[2]Tomsk State University of Control Systems and Radioelectronics, 634050 Tomsk, Russia

Abstract. This paper demonstrates the outcome of experimental studies on thermoelectric characteristics of thermocouples. Measurements were carried out using different types of thermocouples; each studied alone then investigated when they are connected in parallel, in order to simulate the imperfect nature of the various multi-contact surface of the tested object. The investigated types were Chromel-Alumel and Nichrome-Constantan. The thermoelectric characteristics were measured at the temperature range 160° – 400° Celsius, in order to identify the best operating temperature for the hot electrode. Furthermore, the load resistance is another important factor that has been investigated and therefore varied from 1Ω to 10kΩ, in order to determine its effect on the electrical characteristics of thermocouples. Accordingly, these characteristics will help defining the requirements for an optimum thermoelectric testing.

1 Introduction

The thermoelectric testing method is widely used in non-destructive testing of metals and alloys [1-3]. However, unreliable results are often obtained due to the heterogeneity and roughness of the surface, which correspondingly leads to a multipoint contact [4, 5]. The impact of the surface quality has been investigated and the results are presented later in this paper. Multipoint contact results in equivalent circuits of parallel connection of several thermocouples that have different characteristics, due to local fluctuations of the chemical composition of the metal or alloy [6, 7].

In order to take into account the impact of the resulted thermocouples on the inspection results, it is necessary to study their characteristics. According to the carried out literature reviews, the study of thermocouples' characteristics has not been conducted until today. Taking account of these characteristics and the degree of influence of the above factors will significantly improve the technical characteristics of the thermoelectric inspection devices.

2 Experimental investigation

In this study, the characteristics of different parallel thermocouples has been investigated in the range of 160°C - 400°C. Characteristics of the thermocouples, Chromel-Alumel and

* Corresponding author: demo_092@icloud.com

© The Authors, published by EDP Sciences. This is an open access article distributed under the terms of the Creative Commons Attribution License 4.0 (http://creativecommons.org/licenses/by/4.0/).

Nichrome-Constantan have been investigated, when each is alone and when both are connected in parallel.

A block diagram of the experimental setup is shown in Figure 1(a) and it includes:

1) A heat chamber, T;
2) Thermocouples of different types, each consists of hot and cold electrodes;
3) Load Resistance, R_L (of nominal values: 1Ω, 5.2Ω, 10Ω, 100Ω, 1kΩ, 10kΩ);
4) Measuring instruments (voltmeter and ammeter).

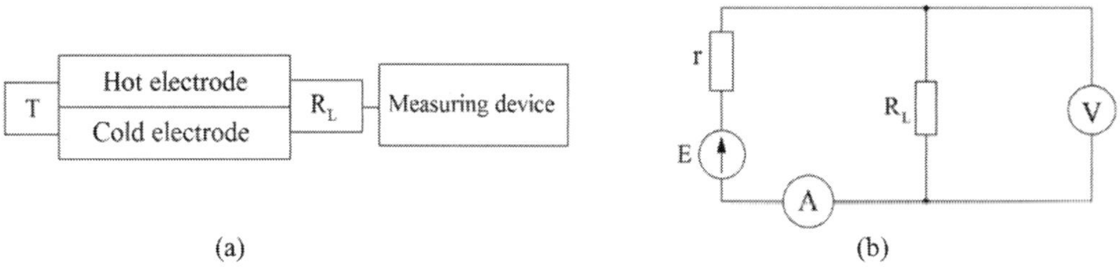

Fig. 1. The experimental setup (a) Block diagram (b) Schematic diagram, where E is the thermoelectric power source, r is internal resistance and R_L is the load resistance.

The thermocouple is presented in the schematic diagram of the experimental setup, which is shown in Figure 1(b), as a voltage source E with an internal resistance r. The thermocouples Nichrome-Constantan and Chromel-Alumel were alternately put in a heat chamber at a varied temperature from 160°C to 400°C. The voltage drop across the load resistance R_L was carefully measured.

Measurements were performed in single and parallel connections of thermocouples, along with the load resistance R_L. Calculations of the thermoelectric power, the internal resistance r, the flowing current in the circuit I, and the maximum electric power P, were all carried out later according to the formulas shown below.

2.1 Calculation of the thermocouple internal resistance, r

According to the second Kirchhoff's law [8], the value of EMF in the circuit is equal to the sum of the voltage drops across the resistors. This means:

$$E = V_r + V_L = Ir + IR_L; \tag{1}$$

where E is the thermoelectric power, V_r is the voltage drop across the internal resistance r, V_L is the load voltage, R_L is the load resistance and I is the current flowing across R_L.

By manipulating expression (1), we obtain:

$$r = \frac{E - IR_L}{I} \tag{2}$$

However, E is considered maximum under the condition of open-circuit test:

$$E = V_{max} = V_{OC}; \tag{3}$$

where E_{max} is the maximum measured thermoelectric power obtained and V_{OC} is open-circuit voltage load.

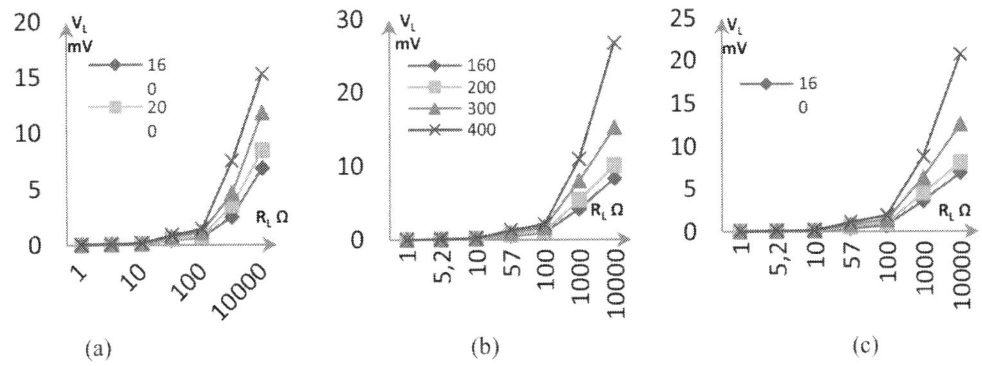

Fig. 2. The dependence curves of the load voltage on the load resistance, where each curve represents a different temperature (160°C, 200°C, 300°C, 400°C). (a) Chromel-Alumel, (b) Nichrome-Constantan, (c) Chromel-Alumel in parallel connection with Nichrome-Constantan.

The results of the voltage measured across the load, with each temperature condition, are presented in Figure 2 as it shows the investigated characteristics of Chromel-Alumel, Nichrome-Constantan and the case for their connection together in parallel.

As shown in the graphs, as the load resistance increases R_L, the load voltage V_L also increases, until it approaches the maximum value of thermoelectric power, at which the open-circuit test is carried out. Upon the increase in temperature, the value of V_L increases as well, due to the increase of thermoelectric power, which is proportional to temperature.

The characteristics of the calculated thermocouple internal resistance have been also investigated and it was found out that as the load resistance increases, the value of the internal resistance increases proportionally. Moreover, it has been observed that the internal resistance, in case of parallel connection of thermocouples, is an equivalent average value of both internal resistance curves of each single thermocouple when investigated alone.

2.2 Calculation of the electric power of thermocouples, P and P_{max}

Power is determined by the expression [9]:

$$P = I^2 R_L, \tag{4}$$

where P stands for power.

From Figure 1(b) we find the current in the circuit: $I = \frac{E}{R_L + r}$ (5)

By substituting the values from formula (4) into formula (5), we have:

$$P = \frac{E^2 \times R_L}{(R_L + r)^2} \tag{6}$$

To find the global maximum point, we take the derivative of P with respect to R_L

$$\frac{dP}{dR_L} = \frac{E^2 (R_L + r)^2 - 2E^2 R_L (R_L + r)}{(R_L + r)^4} \tag{7}$$

After the unification of the general terms we get:

$$\frac{dP}{dR_L} = E^2 \frac{(R_L^2 - r^2)}{(R_L + r)^4} \tag{8}$$

E is never be equal to 0 and r can never be negative. This means the only left solution is:

$$(R_L^2 - r^2) = 0 \tag{9}$$

This means the absolute maximum value of P is always at $r = R_L$. And as shown in Figure 3, the experimental results are confirmed by the calculations presented above. The maximum electric power is consistently observed when $r = R_L$.

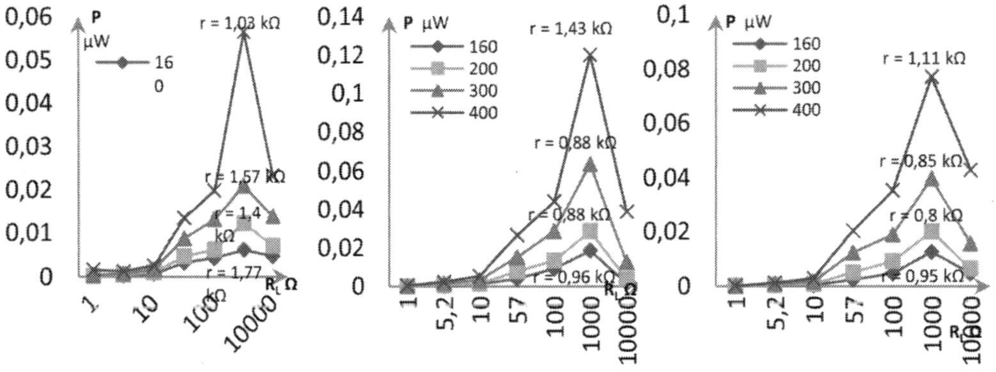

(a) (b) (c)

Fig. 3. The thermocouple power versus load resistance at temperatures (160 - 400)°C (a) Chromel-Alumel, (b) Nichrome-Constantan, (c) Their parallel connection.

3 Conclusions

Within this article, an investigation of the electrical characteristics of two thermocouples in parallel connection was conducted, in order to simulate and study the effect of multi-point contact between the sensing thermocouple electrodes and the inspected sample in thermoelectric inspection devices.

It has become clear from the conducted research that thermocouples, when connected in parallel, values of the internal resistance, electric power and thermoelectric power, are steadily equivalent to the resulted average values from thermocouples Chromel-Alumel and Nichrome-Constantan when each connected alone.

It has also been found experimentally that the most optimal load resistance for thermocouples is 1kΩ. This resistance satisfies the condition $r = R_L$, and consequently when thermocouple electric power is greatest. The choice of this load resistance would improve the accuracy of measurement of the thermoelectric inspection devices.

References

1. C. Stuart, J Test Eval **15**, 224 (1987)
2. P. E. Mix, *Introduction to Nondestructive Testing* (John Wiley & Sons, New Jersey, 2005)
3. A.I. Soldatov et al., *SIBCON*, 7491869 (2016)
4. A.A. Soldatov et al., RUSS J NONDESTRUCT+ **48**, 184 (2015)
5. A.I. Soldatov et al., KEY ENG MAT **685**, 310 (2016)
6. A.I. Soldatov et al., *MEACS*, 7414922 (2015)
7. J. Hu, P.B. Nagy, Appl Phys Lett **73**, 467 (1998)
8. T.H. Davies, Mech Mach Theory **16**, 171 (1981)
9. T. Croft et al., *American Electricians' Handbook* (McGraw-Hill, New York, 2009)

Intersubjective management in aerospace engineering

Mariam Arpentieva[1,*], *Olga* Duvalina[1], and *Irina* Gorelova[2]

[1]Tsiolkovskiy Kaluga State University, 248023 Kaluga, Russia
[2]Russian Academy of National Economy and Public Administration under the President of the Russian Federation, 400131 Volgograd, Russia

Abstract. This article presents a postnonclassical approach to create the science of management processes organization in a developing society, the focus of which is "the man of culture", i.e. the man, not just adhering to cultural norms, but also creating new concepts and products of culture. This science is proposed to be called Evergetics. The purpose of the study is the analysis science of management processes organization in a developing aerospace engineering and other industrial areas of society. The authors describe the main aspects and procedures evergetics management in aerospace engineering. They uses the comparison method, compares classical and modern approaches and technologies of management. In evergetics management model each member of society or organization is interested in augmenting his cultural heritage he is producing, which entails a raise of stability in process of engineering actions and a raise cultural potential of the society as a whole and, as a consequence, an increase in the proportion of moral and ethical managerial decisions and corresponding to them benevolent actions in organizational life. Summarize the article's main findings, authors may in some main conclusions about necessity evergetics model and intersubjective technologies in the creation and development of aerospace engineering.

1 Introduction

In modern science and practice of management the Central issue is to improve the management in the leading fields of development. One such area is the aerospace engineering. Aerospace engineering - the sphere of high technologies in the design and execution of aerospace engineering and in the management of these processes. Achieving sustainable, stable development in this area, therefore, is the issue of optimization of management technologies. The purpose of the study is an analysis of the prospects and difficulties of implementation of intersubjective technology in the aerospace engineering.

* Corresponding author: mariam_rav@mail.ru

© The Authors, published by EDP Sciences. This is an open access article distributed under the terms of the Creative Commons Attribution License 4.0 (http://creativecommons.org/licenses/by/4.0/).

2 Materials and methods

The main method of research is theoretical analysis of the problems of government and industry (including aerospace) in the modern world. Special attention is paid to the concepts of crowd-management and evergetics, as well as intersubjective technologies, differences from monosubjective technologies.

3 Results

The crisis of aerospace engineering in the modern world and States as whole demonstrates that the classical and post-classical model of management obsolete. The modern model of government and communities is being developed in Evergetics. Evergetics is the modern, postnonclassical science of intersubjective management, qualitatively different from the management of bureaucratic, monosubject [1, 2, 3], which explicitly seeks globalization and mondialisation (merge countries in the universal state with a single control, the exception is «public opinion» from the Department of state) and, less explicitly, corporate (split governance within each country and throughout the world under the «power of crowds» transfer of control «public opinion»). Evergetics, in contrast to the classical and non-classical theories, guidelines that serve as corporate, suggests a number of changes.

Evergetics emphasizes cooperation. This cooperation based on the opinions of all parties concerned in a particular decision of people. The essential movable clustering of communities and systems, which traditionally controls the «monolith system» (bureaucracy), restoring relationships «horizontal», need cooperation at different levels and functions of the agencies and organizations. Evergetics form cooperation and mutual support relationship. Evergetics involves a variety of specialists and groups of specialists («stakeholders») from different agencies for the solution of specific problems of the aerospace engineering and neighboring spheres.

This concept and it's technologies stimulates the strengthening of local governance (grassroots level). It is necessary to organize the cooperation of professionals and non-professionals in aerospace engineering. It is important to organize the cooperation of the government, public and business structures as the heterogeneous actors systems. Significant efforts should be made to stimulate processes of self-development and mutual development of economic and other systems: different groups and subgroups of the specialists and non-specialists in aerospace engineering. This is a complex, heterogeneous, lowerarchy ("downstream" elements are sources of resources and power to "higher" peoples), developing system, each component which interprets itself in dialogue with others in «the kaleidoscope of situations», more or less important for them, representative government determines Executive.

4 Discussion

Modern technologies of aerospace engineering industry include three options: bureaucracy, hierarchy, creating monoactor / «monolithic» control systems are widely used, demonstrating the inability to solve problems related to the management of such complex systems as man, society, state. Lowerarchy creating multifactorial management system can be used to solve problems that are difficult or impossible to solve by a single actor «monolithic system».

Corporate or crowd-management supporting quasi-multi-actor system and simulates the intersubjectivity of making and implementing decisions, are used by the bureaucracy as an attempt to «change without changing» [3, 4, 5]. The technology multi-actor / intersubjective management is radically different from the crowd-technologies (table 1).

Evergetics is not based on «artificial» order, supported Monoactor management systems, but the natural order. In traditional management, we are talking about deterministic systems: supported by artificial order, the reliance of managers on legal standards and destruction of legal norms (corruption, bureaucracy), formal hierarchy, proprietary, monologic information exchange and single-handedly solves the problem of the leader. People is a homogenous «herd» or «cattle».

The consumerism as «the ideology of the gangster» stimulates the relations of type of slavery (the involution of humanity). «Universal recipes» of the classical management consideration of situations in all cases cause the system to collapse, the reduction of energy and information. Evergetics involves the management of self-developing systems: the head does not interfere with the natural order: unstructured group with the free, outwardly chaotic, Dialogic information exchange and collective decision-making.

Table 1. Technology intersubjective and crowd management.

Crowd-Technology	Technology Intersubjective Management
Professionals are the members of the «smart crowd». They can participate in the isolation and resolution of problems of different levels, self-organization depending on the need and the degree of involvement and expertise in the matter under discussion. The main question – treatment discussions of crowd problems (processing of information)	Professionals and nonprofessionals as (heterogeneous) actors involved in taking the decision directly relating to their problems, the main issue is the involvement of actors in the process of discussing and implementing decisions (personal responsibility)
The decision taken by the head (quasy-multi-actor technology, simulation of multiactivity /intersubjectivity); the use of mechanisms of total public control, involvement of citizens in relationships in which they are subjects (actors) and objects	The decision implement actors (multi-actor, intersubjective technology) the involvement of citizens in the relations concerning the recognition of subjectivity and subjectivity, the ability to organize themselves
The power is fenced off from citizens with «system agents» for the processing the data of «public opinion» and other means to develop relations of control, increasingly destroying the boundaries of the individual and the organization, as well as prevent it from moral and legal imperatives of culture	The government establishes citizens of the real relations of cooperation, it is transparent, but does not require transparency from individuals and organizations, respecting their boundaries (within the marked moral and legal imperatives of culture)

People and situations are different, the mutual aid as the relationship of social service to allow humanity to evolve, there are no common recipes for all occasions, the discussion about each specific situation leads to an increase of energy and information.

The main change processes, therefore, involve:

- the growth of information and energy (the effect of polylogue), other systems optimizations;
- restoration and development of culture as a system of moral and legal standards, harmony, inclusion and responsibility, transparency and respect (social and human capitals);
- the activation of man, society, state, clustering and self-organization of society (actors and situations);
- sustainable development of human society, the state, anti-crisis effects (smoothing and removal of the crises).

Economic and other system of the aerospace engineering – changing process in which a Manager must account not only for «strong» and «weak signals», including the opinion and life of the «little» people. So instead, to achieve the ideal of «diligence» as «learned helplessness», it is important and necessary to maintain the willingness to give / sacrifice, respect people and government of the dignity of each other. Evergetics suggests that management reform and modernization of its structure includes a number of points:

• rethinking the role of the importance of human and social capital in the development of the state, society and its management in aerospace engineering and other spheres ;

• the formation and application of management technologies, taking into account the «human factor» in aerospace engineering;

• the understanding of the importance of social partnership and cooperation of the «center» and periphery, entrepreneurs and consumers, professionals and non-professionals;

• the development and application of technology cooperation /partnership, their activation in aerospace engineering;

• understanding of perspectives of relations of social service and mutual help, moral supports development;

• (trans)formation and application of technology focused on the development of the cultural capital of communities and organizations;

• de-liberalization of relations between the state, society and business in aerospace engineering and other spheres, the violation of moral norms, overcoming social anomie;

• the decision of problems of social exclusion and lack of accountability: the growth of social inequality, unemployment and precarization of professional work in aerospace engineering;

• the solution to the problems of sustainable development: crises of individual and organizational development (downshifting, professional and organizational collapse);

• the deformalization and dissimulation of state support for its citizens and businesses, to fight corruption at the local level, the periphery of the (moral and legal);

• the creation of mechanisms to support socially responsible business and to encourage the participation of individuals and groups in the management of (moral and legal);

• the introduction of alternative forms of conflict resolution/decision-making (mediation) in health care, education, policing and other fields [1, 4, 5, 6].

The basic principles of the reforms are as follows:

1) the ideology of the «moral fact that profitable,» ignoring the problems and opposition as the suppression of the riots are the society and the country to collapse;

2) the life – changing process that requires attention not only to the «strong» and «weak signals», including the opinion and life of the «little» people;

3) instead of «learnt helplessness», it is necessary to maintain and develop the willingness to give / sacrifice, respect people and government of the dignity of each other.

The mood on the ideals of social service:

1) the reconstruction problem of optimization and development of human communities and the moral foundations of their functioning, orientation to the strategic «benefits»;

2) the required acts of mercy and charity , as well as acts of joint activities of different organizations and people in search of moral forms of governance and of human life, the organization of the state;

3) restoration and development of culture (moral and legal regulator of relations of people and communities).

Modern management in aerospace engineering, however, often involves a number of completely opposite of the developmental, modernization, and phenomena:

1) ignoring the dissent, a lack of attention to social and human capital, giving birth to the deprofessionalization, downshifting professionals and managers in aerospace engineering, as well as the turnover rate of organizational collapse;

2) the production of disposable and/or known to be harmful to human, his mental and physical health goods and services, the crowding out of honest manufacturers and the exploitation of the baser instincts of the «crowd»;

3) the society is moving towards the ideology of consumption, increasing consumerism, more and more clearly understood neo-slavery nature of the relationship in many modern organizations, including the aerospace engineering organizations, the rejection of the innovation or failure.

5 Conclusions

This article presents a postnonclassical approach to create the science of management processes organization in a developing society, the focus of which is "the man of culture" [7]. The lowrarchy as the management of «bottom-up», social service and interaction based on the principles of the gift exchange complementing market, distribution relationships are hard to «restore» its position in the society, and it is organizations [5]. Modern manager in aerospace engineering develops an understanding that the economy in General and manufacturing in particular is not reduced to «business», innovation and change of technological structures puts the center of production and other forms of relationships of the people themselves, social and human capital. It is important to understand that anti-crisis measures – developmental activities: therefore, the restoration of culture, including the moral foundations of the relationship – the main condition of success. Summing up, we note that the results of research in the field of intersubjective management technologies can and should be used in the optimization of socio-political and economic relations in aerospace engineering, in the transition from repressive management to the co-creative artistic and parity relationships [6].

References

1. M. Arpentieva, *Problems of control and modeling in complex systems. Proceedings of the XVIII International conference,* 39 (2016)
2. V.A Wittih., T.V., Moiseeva, P.O. Skobelev, Ontologiya proektirovaniya [Ontology engineering] **2(8)**, 20 (2013)
3. M. Arpentieva, *Proceedings of the XVII International conference «Problems of control and modeling in complex systems», Samara*, 174 (2015)
4. Wallerstein I. *After liberalism* (New Press, New York, 1995)
5. V.A.Vittikh. Group Decis Negot **24**, 67 (2015)
6. V.A.Vittikh, Group Decis Negot **24**, 949 (2015)
7. V.A.Vittikh, Group Decis Negot **24**, 825 (2015)

MATEC Web of Conferences **102**, 01003 (2017) DOI: 10.1051/matecconf/201710201003
V International Forum for Young Scientists "Space Engineering"

Analysis of hydrogenated zirconium alloys irradiated with gamma - rays

Askar Askhatov[1,*], *Vitaliy* Larionov[1], and *Victor* Kudiyarov[1]

[1]Tomsk Polytechnic University, 634050 Tomsk, Russia

Abstract. The paper represents the investigations concerning the geometrical size effect of hydrogenated zirconium alloys (Zr-1Ni-H) during gamma-ray irradiation on the amount of energy absorbed. The results have shown that the less the cross-sectional dimensions of the sample or product is, the less energy is absorbed. The paper provides theoretical calculations. The zirconium sample with a cross-section of 2.8x2.8 cm absorbs 30-35% of the energy of the incident gamma-ray flow. The increase in the cross-section of a product up to 28 cm leads to the increase in the absorbed energy by more than 2 times. At the same time, the thickness of the product is constant. This effect is explained by the multiple scattering of gamma-rays. It leads to the nonuniform distribution of defects which can accumulate hydrogen and should be considered when developing the analysis methods. These edge effects are confirmed by the measurement of the thermal electromotive force for the samples of zirconium alloys before hydrogenation and gamma-ray irradiation, and after irradiation.

1 Introduction

At present, the different modifications of known metal are used to obtain new materials for the aerospace industry. The development of technologies and equipment requires the information on the mechanisms and characteristics of physical and chemical processes, the characteristics of the elemental composition and the structural phases of products [1-5]. Most of the products are exploited under ionizing radiation of different nature: electrons, ions, gamma-rays [1, 2, 5-9]. Radiation is used in the continuous and pulsed modes [2, 6, 7]. It is of particular interest to study the effect of radiation on hydrogenated materials [2, 5]. The properties of these materials are of particular interest to the aerospace materials science. In recent years, these problems have become urgent to researchers in different scientific areas [9], since there is a possibility to achieve the deep and controlled restructuring of metals and alloys on the different structure levels [9]. Various physical methods are used for the analysis of these materials [1-3]. The goal of this work is to study the gamma-ray energy accumulated by the zirconium samples of different sizes. The amount of absorbed energy depends both on the material properties and geometrical sizes. To explain this fact, an analytical calculation method is used [10]. The direct analysis of

[*] Corresponding authors: askhatov10@mail.ru

© The Authors, published by EDP Sciences. This is an open access article distributed under the terms of the Creative Commons Attribution License 4.0 (http://creativecommons.org/licenses/by/4.0/).

changes in the properties of irradiated materials was conducted by the thermal electromotive force method [11].

2 Materials and methods

The samples of the Zr 1% Ni alloy were in the form of a 20 x 20 mm rectangle with a thickness of 1 mm and disc-shaped with a radius of 2 cm and a thickness of 2 mm. The sample hydrogenation was conducted using an installation PCI_ «Gas Reaction Controller» by the Sieverts method at the temperatures T = 450 - 600^{0}C and the pressure of 0.66 atm. The hydrogen concentration was varied from C=690 to C=5000 ppm. The absolute hydrogen concentration in the alloy was determined using a RHEN 602 analyzer (LECO). Irradiation with the Co^{60} gamma-rays was conducted using an installation IRT -2000 (TPU) with different doses of gamma-rays up to 2,6.10^{18} γ-sq/cm^2. The thermal electromotive force was measured using a T-3SP device and a V7-78/1 voltmeter. The thermal probe was made of a gold rod with a diameter of 1 mm.

3 Results

The absorption of gamma-rays with an energy of 80 keV was anylized using the Goryachev-Larionov model [10]. The calculations were performed for the following initial data: the geometric sizes of the target sample (y_0, z_0, $x_o = d$), the attenuation coefficient $\mu(E)$, the total cross section of absorption for Compton scattering Σ^h_g and photoeffect Σ^f_g, the total cross section of scattering Σ_s, the integral «effective» probability of X-ray photon survival for Compton scattering $\Lambda_k = \Sigma_s / (\Sigma_s + \Sigma_g^k)$, the integral «effective» probability of photon survival during the process of photoeffect $\Lambda_\Phi = \Sigma_s / (\Sigma_s + \Sigma_g)$, the integral probability of photon absorption $P = 1 - \Lambda_\Phi \Lambda_k$), and the initial intensity for the incident flow of gamma-ray emission is equal to 1. The amount of energy absorbed in terms of the initial flow is given by

$$E_p = g\frac{[1 - A_\infty(y_0, z_0, E_0)]\{1 - \exp[-K(y_0, z_0, E_0)\mu d]\}}{1 + A_\infty(y_0, z_0, E_0)\exp[-K(y_0, z_0, E_0)\mu d]} \qquad (1)$$

The example of calculation by the formula (1) of γ - ray absorption in the samples of different geometric sizes (zirconium) is illustrated in Figure 1.

The maximum absorption of gamma-rays for this energy is observed for the samples with a cross-section of 0.556x0.556 m and more. But, for example, for a target with a cross-section of 2.8x2.8 cm, relative absorption is ~ 40%. The dependence of γ-rays absorbed on the thickness of the target (x) is obvious and increases with increasing x for the samples of any cross-section. The analysis of calculations also shows that the energy is absorbed nonuniformly over the sample area due to intensive leaving γ-rays through the lateral surface of the target sample.

It can be supposed that the defects will be accumulated nonunifromly according to this effect. Therefore, for analyzing the sample by various methods, there is a need to consider this peculiarity when a sensor moves on the surface of the sample irradiated. It should be noted that this peculiarity of the energy absorption during the multiple scattering of gamma-rays should be considered for the small size samples exposed to irradiation and used in the preliminary experiments. In the calculations, the mass attenuation coefficient was taken to be 1.64 (μ / ρ, cm^2/g) for zirconium.

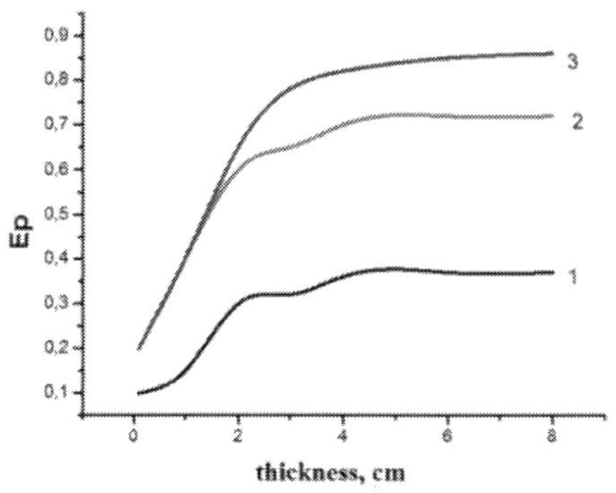

Fig. 1. Absorbed energy (in fractions of the incident flow) versus the sample thickness. (1) the size of cross-section is 2.8x2.8 cm; (2) - 0.28x0.28 m; (3) - 0.33x0.33 m.

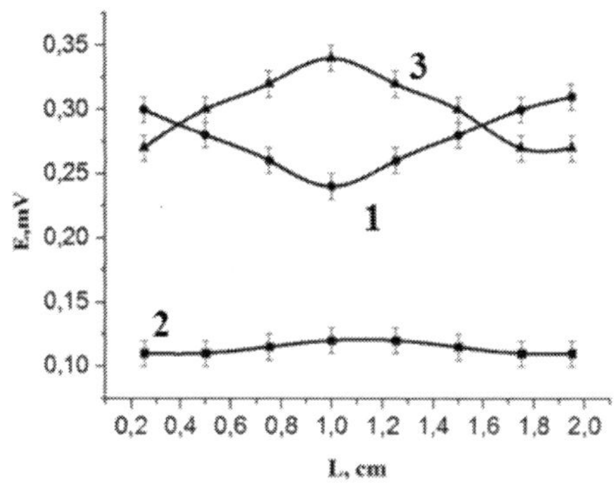

Fig. 2. Thermal electromotive force versus the sensor position on the sample surface. (1) is the hydrogenated sample before gamma-ray irradiation, (2) is the initial commercial Zr-1Ni alloy before hydrogenation, (3) is the irradiated and hydrogenated sample ($C_H = 3000$ ppm).

The value of the thermal electromotive force for the initial sample is constant within the measurement error (Figure 2). This value is increased by more than 3 times, which substantially exceeds the measurement error. Corrosion can occur on the surface of hydrogenated zirconium. It is manifested in the form of a strip. A probe for measuring the thermal electromotive force along and perpendicular to the defect registers the jump of the thermal electromotive force (15-20%) depending on the temperature.

4 Discussion

The highest change of the thermal electromotive force versus the coordinates of the sample was found for the irradiated sample. Edge effects in the hydrogenated sample were caused by the heterogeneity of hydrogenation from the gaseous atmosphere, using the Sieverts method. It should be noted, that this property should be considered in the analysis of physical quantities versus the hydrogen concentration. The integral method for determining the concentration of hydrogen in metals (LECO method) does not consider this peculiarity. The dependence of the thermal electromotive force on the coordinates of the probe for the irradiated sample is reversed with respect to the unirradiated sample. This effect can not be explained by the heterogeneity of hydrogenation. Therefore, the most acceptable explanation is the influence of the edge effect. The edge effect is caused by the intensive leaving gamma-rays through the lateral surface of the sample. This is confirmed by the qualitative calculation shown in Figure 1. Irradiation accelerates the loss of hydrogen by the test material.

This behavior of irradiated materials can be connected with the creation of additional hydrogen vacancies during gamma-ray irradiation. In the irradiated samples, this process is caused mainly by hydrogen diffusion in internal flows which can represent the grain boundaries. In addition, the irradiation of the alloy leads to the change in the concentration of charge carriers, which results in the different behavior of irradiated and unirradiated materials. The measurements conducted in the mutually perpendicular directions along the sample can show the change in the thermal electromotive force up to 15%. This jump in the value of the thermal electromotive force can be explained by the change in the chemical composition, as well as by the deformation of the alloy surface.

5 Conclusions

The theoretical and experimental results have shown that the geometrical sizes of the hydrogenated zirconium alloys Zr-1Ni-H influence on the amount of energy absorbed during gamma-ray irradiation. The calculations demonstrated that the smaller the lateral sizes of the sample or the product were, the less energy was absorbed, regardless the thickness of the sample. The effect is explained by the multiple scattering of gamma-rays. It leads to the nonuniform distribution of defects, which can accumulate hydrogen and should be considered when developing the analysis methods. The presence of edge effects was experimentally confirmed by the measurement of the thermal electromotive force for the zirconium alloy samples before hydrogenation and gamma-ray irradiation and after irradiation. The factor of heterogeneous hydrogenation of samples by the Sieverts method was considered in the measurements of edge effects. The value of thermal electromotive force is changed by more than 1.6 times and depends on the amount of hydrogen injected into the target, the target size and the coordinates of the probe position on the target body.

References

1. K. Oura, V.G. Lifshits, A.A. Saranin, *Vvedenie v fiziku poverkhnosti* (Nauka, Moscow, 2006) (in Russian)
2. I.P. Chernov, S.V. Ivanova, M.Kh Krening, N. Koval', V.V. Larionov, A.M. Lider, N.S. Pushilina, E.N. Stepanova, O.M. Stepanova, Yu.P. Cherdantsev, Technical Physics **57**, 392 (2012)
3. V.V. Ovchinnikov, *UFN* **178**, 9 (2008)
4. E.A. Denisov, T.N. Kompaniets, *ZhTF* **71**, 2 (2001)
5. V.V. Larionov, N.N. Nikitenkov, Yu. I. Tyurin, Technical Physics **61**, 5 (2016)
6. I.P. Chernov, A.S. Rusetsky, D.N. Krasnov,V.V. Larionov, T.I. Sigfusson, Yu.I. Tyurin, Journal of Engineering Thermophysics **20**, 4 (2011)
7. Yu.S. Nechaev, *UFN* **178**, 7 (2008)
8. J. Als-Nielsen, D. McMorrow, *Elements of Modern X-Ray Physics* (Wiley, New York, 2001)
9. A.V. Gapontsev, V.V. Kondratyev, *UFN* **173**, 10 (2003)
10. B.V. Goriachev, V.V. Larionov, S.B. Mogilnitskii, B.A. Saveliev, Atomic energy **62**, 5 (1987) (in Russian)
11. A.M. Lider, V.V. Larionov, M. Kroening, V.N. Kudiiarov IOP Conf. Ser.: Mater. Scien. and Engin **132**, 1 (2016)

MATEC Web of Conferences **102**, 01004 (2017) DOI: 10.1051/matecconf/201710201004
V International Forum for Young Scientists "Space Engineering"

Mathematical model of temperature field distribution in thin plates during polishing with a free abrasive

Alex Avilov[1,*], *Vladimir* Syrovatka[1], *Rustam* Hubiev[1], and *Olga* Baryshnikova[1]

[1]Don State Technical University, 344000 Rostov-on-Don, Russia

Abstract. The purpose of this paper is to estimate the dynamic characteristics of the heating process of thin plates during polishing with a free abrasive. A mathematical model of the temperature field distribution in space and time according to the plate thickness is based on Lagrange equation of the second kind in the thermodynamics of irreversible processes (variation principle Bio). The research results of thermo elasticity of thin plates (membranes) will allow to correct the modes of polishing with a free abrasive to receive the exact reflecting surfaces of satellites reflector, to increase temperature stability and the ability of radio signal reflection, satellite precision guidance. Calculations of temperature fields in thin plates of different thicknesses (membranes) is held in the Excel, a graphical characteristics of temperature fields in thin plates (membranes) show non-linearity of temperature distribution according to the thickness of thin plates (membranes).

1 Introduction

In the aerospace industry circular membranes are used as reflectors of optic-electronic devices for orientation and celestial navigation of spacecraft and satellites. The reflector must have sufficient accuracy of the reflecting surface, which affects the pointing accuracy, temperature stability and radio-reflection ability. Modern technology and equipment used in polishing of thin plates (membranes) allow to receive precision of surface machining, commensurate with the fractions of light wavelength. Membranes are processed by domestic polishing-lapping machine 3PD320 and also by machines OptoTech foreign production using abrasive pastes as a free abrasive [1]. But the process of thin steel plates (membranes) polishing is accompanied by a considerable friction, the uneven distribution of heat flow in the plate material, the fluctuations of the technological system elements, the occurrence of thermo elasticity effects, the substantial depth of the broken layer and other phenomena that affect the quality of thin plates (membranes) surface [2]. As a result of these reasons, about 50% of the treated plates do not meet the required parameters of the surface quality: roughness parameters, accuracy of the surface geometrical shape, deviation from flatness etc. Production reject occurs very often – plate warping. The manufacturer

*Corresponding author: av100982@ramber.ru

© The Authors, published by EDP Sciences. This is an open access article distributed under the terms of the Creative Commons Attribution License 4.0 (http://creativecommons.org/licenses/by/4.0/).

incurs significant material and time costs (expenses), because the polishing operation takes approximately 30 minutes and it is accompanied by a significant consumption of abrasive material. The goal of theoretical studies is the analysis of thermal processes occurring in polishing of thin plates (membranes) with a free abrasive.

2 A mathematical model of the temperature field in thin plates (membranes)

To assess the dynamic characteristics of the heating process in the plate with thickness δ - it is necessary to determine the speed of the thermal wave propagation, for which the variation principle Bio in thermodynamics of irreversible processes has the form [3]:

$$\frac{\partial U}{\partial \alpha_k} + \frac{\partial \Phi}{\partial \dot{\alpha}_k} = F_k, \tag{1}$$

where: U - heat capacity (the temperature elastic potential Bio), an analogue of the potential function in mechanics; Φ -the potential of scattering, analogue of the dissipative function in mechanics; F_k - thermal mechanical strength, the analogue of the external force in mechanics; α_k - the generalized coordinate.

The distribution of temperature field in space and in time according to the plate thickness is approximated by the curve of the second order [3]:

$$T(z,\tau) = T_1(\tau)\cdot\left(1-\frac{z}{\delta}\right)^2, \tag{2}$$

where $T(z,\tau)$ - the distribution of temperature field in time according to the thickness of the plate 0...z...δ; $T_1(\tau)$ - the distribution of temperature field in time; τ - time non-stationary process.

Heat capacity is defined as follows [3]:

$$U = \frac{1}{2}c\rho\int_0^\delta T_1^2(z,\tau)\,dz = \frac{1}{2}c\rho\int_0^\delta T_1^2(\tau)\left(1-\frac{z}{\delta}\right)^4 dz\,, \tag{3}$$

where c - is the specific volumetric heat capacity; ρ - density of the plate material

Thermal displacement H is represented by the ratio [3]:

$$\rho c T_1(z,\tau) = -\frac{\partial H}{\partial z}. \tag{4}$$

Then the expression (4) with (2) has the form

$$H = c\rho T_1(\tau)\left(\frac{z^2}{\delta} - \frac{1}{3}\frac{z^3}{\delta^2} - z + \frac{\delta}{3}\right). \tag{5}$$

To calculate the potential scattering [3] computed the derivative of the magnitude of the thermal drift:

$$\Phi = \frac{1}{2\lambda_1}\int_0^{\delta} \dot{H}^2 dz\,, \tag{6}$$

where: λ_1 - coefficient of heat conductivity of the plate material.

The expression (5) after the differentiation and integrating has the form:

$$\int_0^{\delta} \dot{H}^2 dz = c^2\rho^2 T_1^2(\tau)\dot{\delta}^2\left[\frac{z}{9} - \frac{2z^3}{9\delta^2} + \frac{z^4}{9\delta^3} + \frac{z^5}{5\delta^4} - \frac{2z^6}{9\delta^5} + \frac{4z^7}{63\delta^7}\right]_0^{\delta} = \frac{13\delta}{315}c^2\rho^2 T_1^2(\tau)\dot{\delta}^2 \tag{7}$$

By substituting the expression (7) into (6) and after differentiation finally we have:

$$\frac{\partial\Phi}{\partial\dot{\delta}} = \frac{13}{315}\frac{c^2\rho^2 T_1^2(\tau)}{\lambda_1}\dot{\delta}\cdot\delta \tag{8}$$

Generalized thermal force F is represented as a ratio of the variations:

$$F = T(\tau)\cdot\frac{\delta H}{\delta z}|_{z=0}\,. \tag{9}$$

Or taking into account the expression (5)

$$F = \frac{c\rho}{3}T_1^2(\tau). \tag{10}$$

The expression heat capacity (3) after the integration has the form:

$$U = \frac{1}{2}c\rho T_1^2(z,\tau)\left(z - \frac{2z^2}{\delta} + \frac{2z^3}{\delta^2} - \frac{z^4}{\delta^3} + \frac{z^5}{5\delta^4}\right), \tag{11}$$

then

$$\left.\frac{\partial U}{\partial\delta}\right|_{z=0} = \frac{c\rho}{10}\cdot T_1^2(\tau). \tag{12}$$

Substituting the obtained expressions (8), (10), (12) the Lagrange equation of the second kind (1), we have:

$$\frac{d\delta}{dt}\cdot\delta = \frac{7}{30}\cdot\frac{315}{13}\cdot\frac{\lambda_1}{c\rho} = 5{,}65\cdot\frac{\lambda_1}{c\rho}. \tag{13}$$

After integration and transformation of the expression (13) finally we have received the form: $\delta = 3{,}36\cdot\sqrt{\frac{t\cdot\lambda_1}{c\cdot\rho}}$.

From the above calculation we have the following form:

$$t = \frac{c\rho\delta^2}{11{,}29\lambda_1}, \tag{14}$$

where: *t*- is the time-setting process temperature.

In Figure 1 the results of the researches are presented, the graphic characteristics of the temperature field in plates of various thickness have been received. The design results of time for plate loss stability, according to the form (14), i.e. the time, after which the warping of the plate (relaxation-setting process temperature) begins, is represented in figure 2.

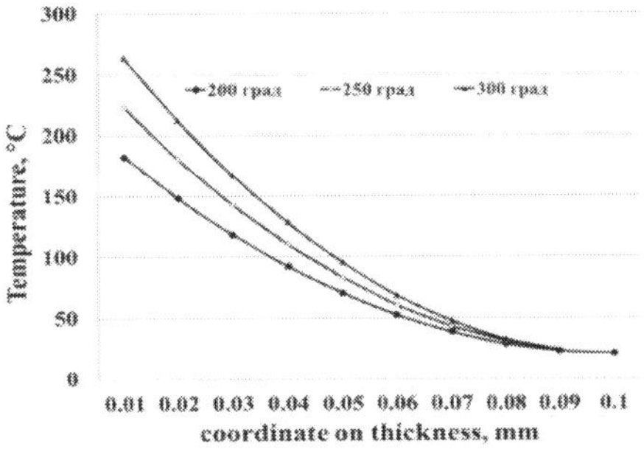

Fig. 1. Temperature field of the plate is h=0.1 mm.

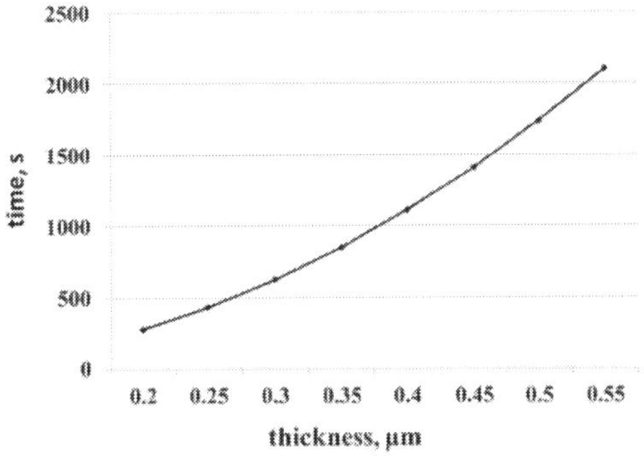

Fig. 2. The relaxation time of process of heating of the plate.

3 Conclusions

According to the results of theoretical researches we can draw the following conclusions:
1. The temperature field varies nonlinearly across the thickness of a thin plate, i.e. the heat flux in the plate is distributed unevenly, causing the appearance of the temperature elastic effects and, hence, loss stability and warping.
2. The relaxation time of heating process of a thin plate depends on the thickness of a thin plate, has a nonlinear character and increases with increasing of the plate thickness.
3. To eliminate the warping process of thin plates it is necessary to adjust the technological process of thin plates polishing, including the modes of polishing to reduce operating temperatures when polishing.

References

1. V.G. Miroshnichenko. STIN. J. **5**, 12 (2009)
2. A.M. Zenkour. Quart. J.of Mech. and Appl. Math. **57**, 205 (2004)
3. A. D. Kovalenko. Fundamental bases of temperature elastic (Scientific thought, Kiev, 1970)

Gating system design for the space device case using T-Flex CAD

Munkhe-Zul Ayusheev[1,*], and *Tamara* Kostyuchenko[1]

[1]Tomsk Polytechnic University, 634050 Tomsk, Russia

Abstract. The judicious selection of gating system for the consumable pattern takes a lot of time, labour and other significant resources. The modern design technologies provide quick and effective ways for gating system calculation and casting process simulation. Gating system modeling allows estimating different kinds of defects which can occur at the developing stage of casting process. Moreover, it is possible to modify the whole gating system configuration if some parameters are changed. Analyzing these data and modifying the gating system characteristics high quality of castings can be achieved.

1 Introduction

Casting is a very crucial manufacturing process for the production of space devices because many of them have cast details. Casting is one of the most economic and efficient method. But, design calculation of the gating system with necessary characteristics to cast product takes a lot of time [1-9].

The modern parametric T-Flex CAD [5-6] .system which was developed by Top Systems Ltd. (Moscow, Russia) can carry out a number of problems and accelerate the process of the gating system design significantly. T-Flex CAD affords an opportunity to modify the whole configuration changing only one or several gating system geometrics, at the same time providing calculations of the necessary parameters depending on the model geometry.

The main results of the gating system design calculation for the space device case using T-Flex CAD are provided. Figure 1 shows 3D model of the developed device case.

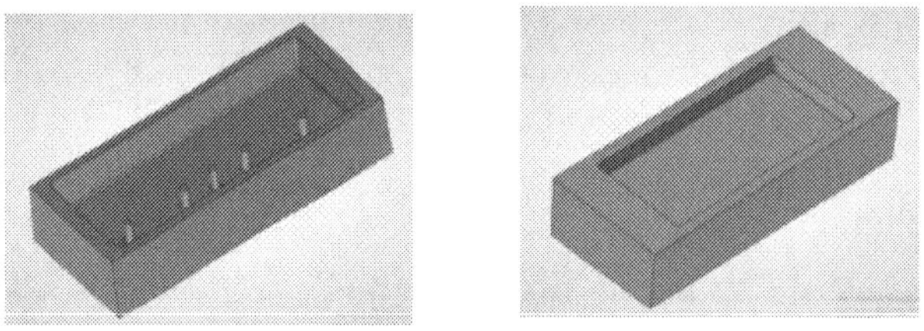

Fig. 1. 3D model of the developed device case.

*Corresponding author: muncko94@mail.ru

© The Authors, published by EDP Sciences. This is an open access article distributed under the terms of the Creative Commons Attribution License 4.0 (http://creativecommons.org/licenses/by/4.0/).

2 Materials and methods

Initial data for the casting:
- It should be a sand-mold casting;
- Device case material – aluminium alloy AK7;
- Device case dimensional specifications 820×350×170 mm;
- Pure casting mass – 50 kg.

According to standard methods for the gating system design we should follow by the next steps [7-9]:

1. Choosing the casting position in a mould, mould joint, casting head height and type of the gating system (pouring temperature for the aluminium alloy AK7 is equal to 710 °C).

2. The typical configuration L-1A of the gating system was chosen [7]. Further, the length of runner branch, number of gates, and sprue position on the runner should be determined. For the device case the length of runner branch should be L_r = 90 cm; number of gates n = 16; sprues will be placed one by one at the each edge of runner.

3. Based on the casting mass and height, gating ratio F_s:F_r:F_g = 1:2:4 [7] is set approximately, where F_s:F_r:F_g – total cross-sectional area of sprues, runners and gates. Sprue diameter will be equal to 1 cm.

According to initial data for the device case (Figure 1), the parametric 3D model of the gating system was created using T-Flex CAD facilities. This model is shown in Figure 2.

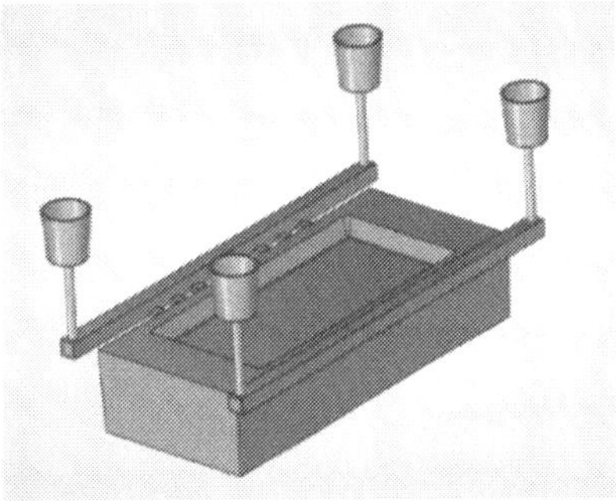

Fig. 2. 3D model of the designed gating system.

The parametric gating system model includes all calculated parameters which depend on the geometrics. Calculation in the T-Flex CAD system is based on data which were introduced in the built-in variable editor.

The cross-sectional area and perimeter at the liquid metal entrance level is defined preliminary. This case, the cross-sectional area is equal to S_c = 646 cm^2 and perimeter P = 435 cm.

The primary hydrostatic head is determined as:

$$H_0 = h_s + h_l, \quad (1)$$

where h_s is the sprue height, h_l is the liquid metal level into the pouring basin. An average head can be calculated by the formula:

$$H_a = H_0 - \frac{h_c}{2}, \quad (2)$$

where h_c is the casting height from the gate level.

The primary and average heads are identical and equal to 24,5 cm as the top gate was chosen, the casting height is equal to zero.

The sprue height is equal to h_s = 16,5 cm. On the assumption of dependence between sprue and basin heights, the height of last is equal to H_b = 10 cm [7]. The liquid metal level in the basin is set to h_l = 8 cm.

For the filling calculation, the average rate of the liquid metal fills the mould cavity is determined by the following formula:

$$v_a = \frac{Q_{max} \cdot \sqrt{H_a}}{S_c \cdot \sqrt{H_0}}, \tag{3}$$

where Q_{max} is the maximum allowable consumption.

The average liquid metal fills the mould cavity is equal to 1,68 cm/sec.

At calculations, it is necessary to keep the filling criterion: the average filling rate has to be more than minimum allowable filling rate.

The vertical walls of the device case which have the thickness about 27 mm will be the most dangerous zones for filling. According to special reference data [7], the minimum filling rate for vertical walls with such thickness (δ = 27 mm) and height (h = 180 mm) will be v_{fmin} = 0,52 cm/s. The calculated average filling rate v_a = 1,68 cm/s is more than minimum allowable rate, therefore, the filling criterion is observed.

The calculated flow rate of liquid metal through the sprue is determined by the monogram at the $d_{l.s.}$= 1 cm, H_0 = 24,5 cm, L_r = 90 cm, n = 16, F_s:F_r:F_g = 1:2:4 and t_p = 710°C, where $d_{l.s.}$ is the sprue diameter in the lower section, t_p is the pouring temperature. The calculated flow rate of liquid metal is equal to v_s = 1,5 m/s. Such rate is less than maximum allowable rate for the sprue with diameter 10 mm [7].

After the diameter setting at the lower section it is necessary to calculate the diameter at the upper section of sprue by the next formula:

$$d_{u.s} = d_{l.s} + 1,5\frac{h_s}{100}. \tag{4}$$

The number of sprues is determined from the total cross-sectional area of sprues. Accepting the number of sprues is four; diameter of each sprue will be 1 cm at the lower section and 1,3 cm at the upper section.

Calculation of the actual flow rate of liquid metal through the runner is determined by the following formula:

$$v_r = v_s \cdot \frac{F_s}{F_r}, \tag{5}$$

where v_s is the flow rate of liquid metal through the sprue.

The actual flow rate of liquid metal through the runner is equal to 75 cm/s.

The runner sizes issue from the runner total cross-sectional area. Consequently, it is possible to accept two runners with identical sizes in cross-sectional geometry for the developing gating system: a = 2,5 cm, b = 2,5 cm: the runner length was determined above.

As $v_r < v_{r.max}$, the condition of liquid metal passing through the runner will be observed [7].

Taking into account that F_g= 4·F_s and accepted number of gates is n = 16, the area of choke can be found as $f_g = \frac{F_g}{n}$. The choke area is equal to f_g=1,8cm^2.

The setting gates thickness is equal to $\delta_g = 0{,}7$ cm that meet a requirement $\frac{h_r}{\delta_g} \geq 3{,}25$, h_r is the runner height. The choke width can be found as $c = \frac{f_g}{\delta_g}$. And the choke width is equal to $c = 2{,}6$ cm. The actual flow speed in gates can be determined as $v_g = v_s \cdot \frac{F_s}{F_g}$.

The actual flow rate of liquid metal through the gates will be equal to $v_g = 41$ cm/s.

The maximum allowable rate $v_{g.max} = \frac{12}{R_g}$. The maximum allowable rate is equal to $v_{g.max} = 48.7$ cm/s.

As the actual rate is less than maximum allowable rate, passing through the gates is provided.

The distance between the sprue and first gate can be calculated as $L_g = 0{,}147 \cdot h_r \cdot v_r$. This distance will be equal to $L_g = 28$ cm.

The calculated parameters meet the all requirements for the gates. Gates has a uniform distribution at the perimeter keeping the distance between the first gate and sprue.

The volume of the filled basin is calculated by the following formula: $V_b = 1{,}5 \cdot Q_{max}$ and this parameter is equal to 1631 cm^3.

The calculations of all described parameters were carried out by using built-in variable editor in the T-Flex CAD system. Changing of the gating system geometrics all parameters will be re-computed for the new configuration automatically.

3 Conclusions

Using T-Flex CAD, the gating system for the space device case (Figure 1) was completely designed and calculated; 3D model of the device case and gating system are parametrically connected. It gives the possibility to have only one case model to receive alternative case configurations with different sizes and automatically reconstruct the gating system model using the new case geometrics.

Such method of the gating system design significantly accelerates designing process and gives the high quality of casting. Besides, the gating system model can be used for further researches by casting process simulation tools.

References

1. V. Dmitriev, Y. Britova, MATEC Web of Conf. **48**, 01005, (2016)
2. A. Swapnil, S.B. Ambekar Dr., Jaju Int. J. Eng. Res. and Gen. Scien. **2**, 93 (2014)
3. N.A. Kupriyanov, F.A. Simankin, K.K. Manabaev, IOP Conference Series: Materials Science and Engineering **124 (1)**, 012100 (2016).
4. F.A. Simankin, G.R. Ziyakaev, I.V. Kuznetsov, R.E. Lushnikov, IOP Conference Series: Materials Science and Engineering **124 (1)**, 012072 (2016)
5. Yu. Britova, V. Dmitriev, T. Kostyuchenko, IOP Conference Series: Materials Science and Engineering **132(1)**, 012011 (2016)
6. A.V. Taracenko, V.S. Dmitriev, T.G. Kostuchenko, *KORUS*, 83 (2004)

7. N. Galdin, *Litnikovye sistemy dlja otlivok iz legkih splavov* (Mashinostroenie, Moscow, 1978) (in Russian)
8. V. Anjo, R. Khan, J. of Practices and Technologies **23**, 45 (2013)
9. B. Vijaya Ramnath et al. Procedia Materials Science **6,** 1312 (2014)

Decrease uncertainty of measuring small differential signal against large common-mode signal

Pavel Baranov[1,*], *Valeriy* Borikov[1], *Edvard* Tsimbalist[1], and *Bien* Bui Duc[1]

[1]Tomsk Polytechnic University, 634050 Tomsk, Russia

Abstract. Comparators capable of comparing two alternating signals of the similar frequency within a wide dynamic range of frequencies and voltages are lock-in amplifiers with a differential input. The paper presents the methods and means are suggested for reducing the measurements uncertainty conditioned by the finite value of the common-mode rejection ratio in lock-in amplifier with a differential input.

1 Introduction

A comparison of the alternating voltage values is the classical problem of instrumentation [1-2]. A method of measurements by comparison against an actual measure is commonly used providing a comparison of the value to be measured $u_x(t)$ and the value of calibration signal $u_0(t)$ reproduced by measure. The practical implementation of this method including bridge and differentiation circuits requires the availability of high-sensitivity comparators possessing a resolution that, in large part, defines the uncertainty of measurements [3-8].

Comparators capable of comparing two alternating signals of the similar frequency within a wide dynamic range of frequencies and voltages are lock-in amplifiers with a differential input [9-11]. These devices are intended for the identification of a small differential signal of voltage to be compared (up to nanovolt units) against a large common-mode signal at a signal/noise ratio of -100 dB and high impedance (up to tens and hundreds of megaohms) at measurement inputs [12-13].

In this work, the methods and means are suggested for reducing the measurements uncertainty conditioned by the finite value of the common-mode rejection ratio in lock-in amplifier with a differential input.

2 Tracking power supply for differential signal extraction

The output voltage of the simplest lock-in amplifiers with a differential input can be obtained from:

* Corresponding author: bpf@tpu.ru

© The Authors, published by EDP Sciences. This is an open access article distributed under the terms of the Creative Commons Attribution License 4.0 (http://creativecommons.org/licenses/by/4.0/).

$$\Delta U \approx \frac{1}{E}\left(\lim_{T\to\infty} \frac{1}{T}\int_0^T \left[u_x(t) - u_0(t) + \frac{u_x(t)+u_0(t)}{2K_{\mathrm{RR}}(f)} \right] u_{\mathrm{ref}}(t)dt \right), \quad (1)$$

where E is the denominator of synchronous detector, V;

$u_0(t)$, $u_x(t)$ are input voltages to be compared, V;

$u_{\mathrm{ref}}(t)$ is the reference voltage, V;

K_{RR} is the common-mode rejection ratio.

From (1), measurement of the voltage difference ΔU of $u_0(t)$ and $u_x(t)$) is accompanied by the uncertainty that depends upon the phase shift φ between input signals and the finite value of the common-mode rejection ratio K_{RR} of the comparator circuit (subtractor). Also, it depends on incoherence of reference $u_{\mathrm{ref}}(t)$ and detected voltages at the inputs of the synchronous detector that is characterized by the phase shift φ_{ref}. The requirements for the common-mode rejection ratio must be improved, in particular, when checking and calibrating the scaling measurement transducers at (1-10 nV) maximum resolution of the lock-in amplifier within the medium frequency range at the higher level of $10\sqrt{2}$ V of compared voltage dynamic range.

The increase of the common-mode rejection can be implemented by the addition of the voltage follower to the circuit as shown in Figure 1 that provides the tracking power supply of the instrumentation amplifier.

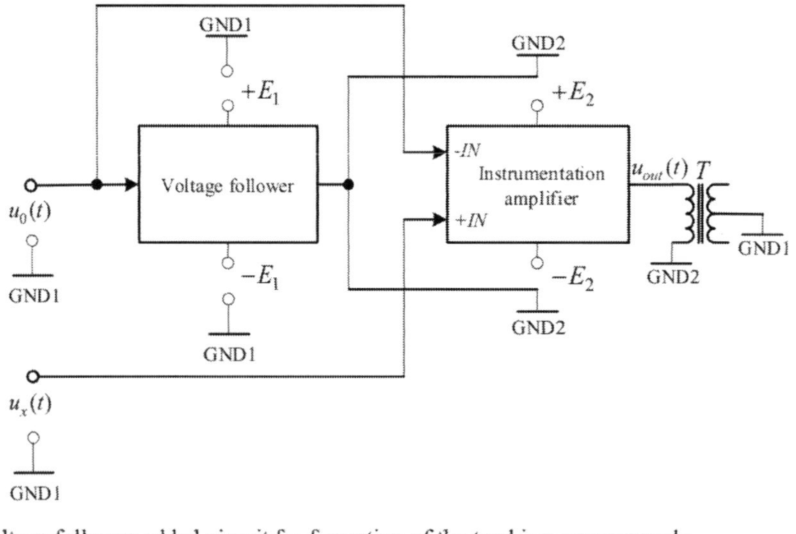

Fig. 1. Voltage follower-added circuit for formation of the tracking power supply.

For this circuit, the efficient common-mode rejection ratio will be increased and comes to

$$K_{\mathrm{RR.ef}}(f) = \frac{K_{\mathrm{RR}}(f)}{1-K_{\mathrm{f}}(f)}, \quad (2)$$

where K_{f} is the transmission factor of the voltage follower.

The common-mode rejection down to 160–180 dB can be achieved within the frequency range of practically inertialess circuits, at easily reachable values of 0,999-0,9999 of the follower transmission as shown in Figure 2. These values provide the stability of the voltage follower transmission factor at different signal sources.

These dependencies prove a certain drawback of this technique used in the wide frequency range: the transmission factor module of the voltage follower and instrumentation amplifier decreases with the increase of frequency resulting in the common-mode rejection ratio of the whole circuit. Therefore, the high rejection of the common-mode signal is observed only in the narrow band of frequencies.

It should be noted that tracking power supply circuit with the use of voltage follower provides also a jump of the input impedance of measuring channels in the wide frequency range, especially u_x channel of the instrumentation amplifier.

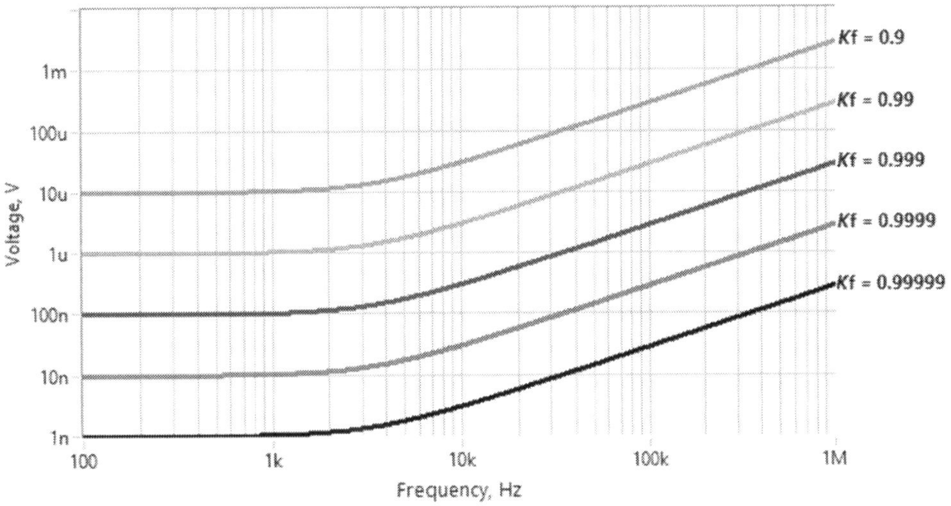

Fig. 2. Output voltage and frequency dependences of PGA207 instrumentation amplifier at different K_f values at common-mode 10 V and zero phase shift of the follower.

3 Sample and hold circuit

In order to achieve the required common-mode rejection, additional the sample and hold circuit can be used in the lock-in amplifier as shown in Figure 3.

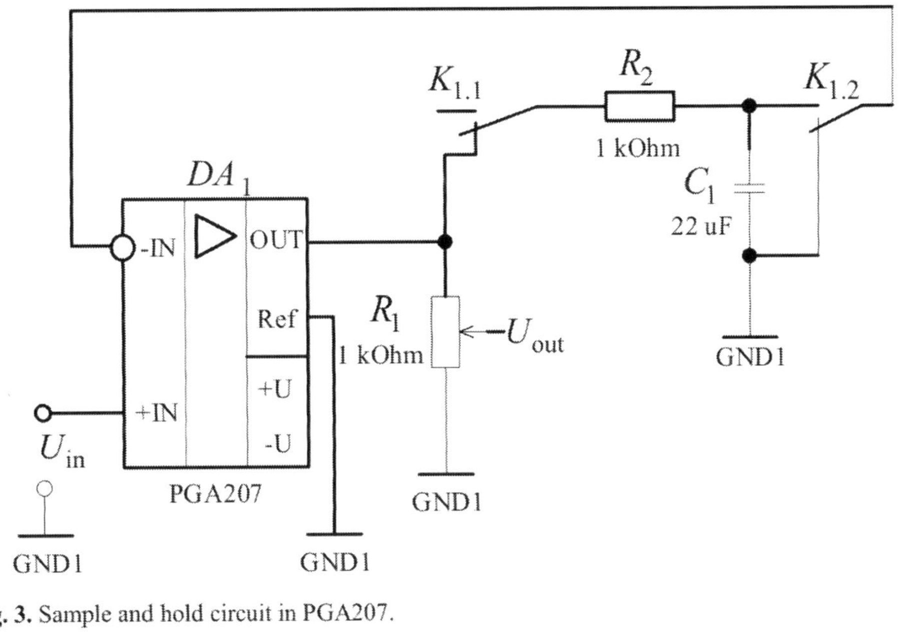

Fig. 3. Sample and hold circuit in PGA207.

With this view, the switch K_1 (e.g. reed switch RES-55) is installed in the connection circuit of the instrumentation amplifier to supply measurement inputs of voltage $u_0(t)$. The discharge common-mode voltage processed by synchronous detector and low-pass filter is stored by condenser C_1 of the sample and hold circuit.

In the mode of measuring the differential signal, voltages $u_0(t)$ and $u_x(t)$ transmit to the measurements inputs of the instrumentation amplifier by switch K_1. At the same time, in the sample and hold circuit the voltage stored in the capacitor is subtracted from the input voltage.

4 Conclusions

The suggested procedures oriented towards the decrease of uncertainty of measuring the small differential signal against a large common-mode signal, meet the requirements of the up-to-date instrumentation.

The technical implementation suggested for the tracking symmetrical power supply for the instrumentation amplifier is based on the voltage follower that allows increasing the common-mode rejection ratio up to 160-180 dB, comparing voltages up to $10\sqrt{2}$ V within the 100 kHz frequency range.

The work has been supported in part by the grant of the RFBR, research project No. 15-08-01007.

References

1. S. Avramov-Zamurovic, G.N. Stenbakken, A.D. Koffman, N.M. Oldham and R.W. Gammon, IEEE Trans. Instrum. Meas. **44,** 904 (1995)
2. L. Callegaro, V. D'Elia, IEEE Trans. Instrum. Meas. **51,** 632 (2002)
3. L. Callegaro D. Serazio, *CPEM*, 222 (2000)
4. J.C. Hsu, and Y. Ku, *CPEM*, 429 (2000)
5. A.V. Rogovyh, N.M. Natalinova, A.S. Spiridonova, A.S. Gordynets, *SIBCON*, 113590 (2015)
6. A. Dolgih, V. Martemyanov, *SIBCON*, 7491844 (2016)
7. V. Martemjanov, A. Dolgih (Ivanova), *SIBCON*, 6693632 (2013)
8. N. Natalinova, N. Ilina, E. Frantcuzskaia, IOP Conf. Series: Materials Science and Engineering, **132**, 012029 (2016)
9. P.F. Baranov, E.I. Tsimbalist, V.N. Borikov, D.G. Soltanova, *IMEKO*, 116100 (2015)
10. P.F. Baranov, E.I. Tsimbalist, V.E. Baranova, *SIBCON*, 7146992 (2015)
11. M.L. Meade, *Lock-in amplifiers: Principles and Applications* (London: Peter Peregrinus Ltd, 1983)
12. S. Fu, A. Sakurai, L. Liu, F. Edman, T. Pullerits, V. Öwall, K. Karki, Rev. Sci. Instrum. **84,** 115101 (2013)
13. M. Min, O. Martens, T. Parve, Measurement **27,** 21 (2000)

Local deformation method for measuring element tension in space deployable structures

Sergey Belov[1,*], *Mikhail* Pavlov[1], *Viktor* Ponomarev[1], and *Sergey* Ponomarev[1]

[1]National Research Tomsk State University, 634050 Tomsk, Russia

Abstract. The article describes the local deformation method to determine the tension of cord and thin membrane elements in space deployable structure as antenna reflector. Possible measuring instrument model, analytical and numerical solutions and experimental results are presented. The boundary effects on measurement results of metallic mesh reflector surface tension are estimated. The study case depicting non-uniform reflector surface tension is considered.

1 Introduction

As mentioned in [1] deployable space structures as antenna reflector are widely used because of their effective deployment within significant coverage and relatively small mass. Nowadays, research work concerning deployable antenna reflector are very actual because of growing communication, monitoring and observation of Earth systems. Some of them could be found in [2-5].

In antenna designing it is crucial not only to control cord element tension, being load-bearing elements, but also reflector metal mesh surface tension. Dynamic characteristics (such as natural frequencies) depend on cord element tension, while reflector surface tension significantly influences its radio – reflective properties. Measuring both cord and reflector surface tension is based on the local deformation method (LDM). This is a quasistatic method for material testing via mechanical enforcement on a solid body surface. In the present work LDM involves the following: tested element (cord or metallic mesh) exposed to given strain. In this case the deformation resistance force is measured, defining the initial tension.

2 Analytical expressions to determine initial cord and membrane tension

Obtained initial cord tension expression to given depth indention and force is applied in the following problem. Cord with cross-section S, initial length L and modulus of elasticity E has initial tension T_0^c and is fixed at the ends. Centrally, load P acts on the cord, bending the cord to corresponding value H. It is necessary to determine the initial tension value in terms of P and H values. Cord elongation could be as following:

* Corresponding author: belovsv@niipmm.tsu.ru

© The Authors, published by EDP Sciences. This is an open access article distributed under the terms of the Creative Commons Attribution License 4.0 (http://creativecommons.org/licenses/by/4.0/).

$$\Delta = \frac{\left(T - T_0^c\right)L}{ES} - \frac{D}{2T^2} \tag{1}$$

where, T – cord tension under load, $D = \int_0^L Q^2 dx$, Q – transverse force in the cord. If in (1) Q is the concentrated load centrally applied, then the expression for cord initial tension is the following:

$$T_0^c = \frac{LP}{4H} - \frac{2H^2 ES}{L^2}. \tag{2}$$

If (2) is divided by S, then cord initial stress is the following: $\sigma_0^c = \dfrac{LP}{4HS} - \dfrac{2H^2 E}{L^2}$.

Analytical expression (3) for circular membrane initial tension is defined by the virtual work principle [6]:

$$\sigma_m^0 = \frac{P_m}{2IH_m \pi}. \tag{3}$$

where, $I = \int_d^b \left[B^2 \left[1 - \frac{1}{\sqrt{1 - H^2 B^2}} \right] - \frac{1}{r} \int_b^r \frac{B^2}{\sqrt{1 + H^2 B^2}} dr \right] r dr$, $B = \dfrac{r\left(B_1 - 2r^2\left(b^2 + d^2\right)\right)}{b^4 - d^4 + 4b^2 d^2 \ln \frac{d}{b}}$,

$B_1 = 4b^2 r^2 \ln \frac{r}{b} + 2b^2\left(d^2 + r^2\right)$, b – circular membrane radius, P_m – centrally uniform membrane lateral load applied on the circle radius d, H_m – maximum membrane depth of indention in the circle of radius d.

3 Experimental results in measuring initial cord and membrane tension

Figures 1 a), b) show experimental units to measure initial cord and membrane tension, respectively, as well as to justify expression (2) – (3).

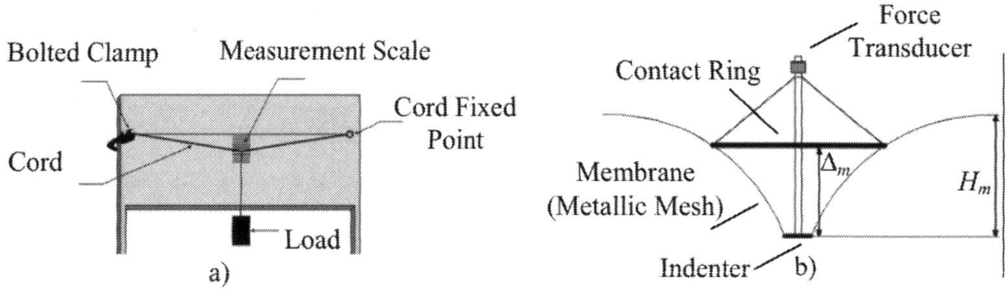

Fig. 1. Diagram of experimental units: a) for cord element b) for membrane element.

3.1 Cord initial tension measurement

Aramide – wired cord with glass core of 0.8mm in diameter and circular cross-section was used in the experiment. If the cord strain is less than 0.3%, then the linear stress-strain dependency exists. In case of 90^0 rotation relative to illustrated position Figure 1a) cord is vertical, while its fixed point is at the top. Load at unfixed end provides cord initial tension T_0^c. After, this end is fixed by bolted clamp and the cord is reversed to initial position with centrally concentrated load *P*. The corresponding cord bending *H* at this load is measured according to the scale which is positioned to the initial cord position without a load. Then, *H* and *P* values are put in (2) which expresses experimental initial cord tension values. These values are compared with respective T_0^c. In the experiment, $T_0^c = 1,2,3$ kgf values were considered. It was shown that the differences between T_0^c and values obtained by (2) are not more than 10%. Thus, the expression (2) is justified.

3.2 Membrane initial tension measurement

To measure the membrane (metallic mesh) initial tension, the strain is set to its normal. On the experimental unit (Figure 1b) the bending depth is controlled by the contact ring. In the experiment the contact ring plane is parallel to the membrane tangent plane. Indenter presses into the membrane until the contact ring touches the membrane surface. At this moment the membrane resistance force is measured by the force transducer. Depth of indention H_m is determined as the Δ_m distance between contact ring plane and indenter edge.

In the experiment the square metallic mesh membrane of 25×25cm was used. Each side was subjected to the distributed load. The side total loads were 0.1; 0.2; 0.3kgf for each experiment, respectively. The inital membrane uniform tensions were $4\cdot10^{-3}$; $8\cdot10^{-3}$; $12\cdot10^{-3}$ kgf/cm. Cardboard contact ring and indenter of $2.5\cdot10^{-3}$kg were positioned in the membrane center. Indenter contact surface is a circle with diameter of 3cm. Loads $P_m=3\cdot10^{-3}$; $4\cdot10^{-3}$; $5\cdot10^{-3}$kgf were imposed on indenter. Thus, the total loads (taking into account mass of indenter) on membrane were $5.5\cdot10^{-3}$; $6.5\cdot10^{-3}$; $7.5\cdot10^{-3}$kgf. Δ_m distance was measured by cathetometer with accuracy of 0.1mm. H_m value was calculated by analytical formula for relevant problems of [7]. Additionally, the same experiments were done for a membrane of 22×22cm with indenter diameter of 1.5cm. In this case, the membrane tesion was varied from $3.8\cdot10^{-3}$ to $15\cdot10^{-3}$kgf.

To calculate initial membrane tension, the finite element method (FEM) was used [7-9]. Square membrane (with fixed edges) of 25×25cm was modeled by shell elements of first order with initial tension option. Its modulus of elasticity is $E_m=2\cdot10^2$kgf/cm^2, Poisson ratio is $\nu=0.29$, material properties are isotropic.

Indenter action was modeled by the forces applied to the nodes in the indenter and membrane contact region of 0.75cm radius .

Based on analytical expression (3), experiments measurement and FEM solutions, depth of indention H_m dependency on σ_m^0 for different load values $P_m=5.5\cdot10^{-3}$; $6.5\cdot10^{-3}$;$7.5\cdot10^{-3}$ kgf were calculated. It was shown that the differences between H_m obtained experimentally are no more than 15% comparable to ones obtained by (3) and FEM solutions. Thus, expression (3) could be justified for membrane initial tension calculation.

In the estimation of boundary membrane closeness effects to the points of measurment, it was shown that if membrane measuring regions are twice more than the contact ring diameter, then membrane boundary effects are insignificant. If these regions are less than the contact ring diameter, the membrane boundary effects are significant (i.e. membrane force resistance increases to indenter load).

It is important to consider the case of membrane biaxial tension with different components of principal stresses σ^0_{xm}, σ^0_{ym}. For this purpose, indenter with elongated shape was used (Figure 2).

Fig. 2. Indenter with elongated shape.

In this case, resistance force depends on depth indention, membrane tension and indenter orientation in the (*x, y)* - plane. Indention was modeled by FEM where contact membrane shell elements with non-uniform tension were used. For above described indenter of length L_{in}, the rounded radius is the half of its width $\rho = H_{in}/2$. Tension along axis x is constant, i.e. σ^0_{xm}=5·10^{-3} kgf/cm. Tension along axis y is varible witin σ^0_{xm}=2·10^{-3} - 12·10^{-3} kgf/cm. Indenter effect was modeled on the basis of given node displacements. It was proved that indenter orientation effect on indention resistance force is more if the indenter is more elongated and bending depth is also more.

4 Conclusions

Obtained results proved the following: a) LDM could be effectively applied to determine initial tension in cord and membrane elements of space deployable structures as antenna reflector; b) analytical expressions (2)-(3) for cord and membrane initial tensions are in a full agreement with experiment and numerical results; c) membrane boundary effects should be considered in the regions where it is less than the contact ring diameter; d) in case of non-uniform membrane tension, elongated indenter could be applied.

This work has been financially supported by Ministry of Education and Science of Russian Federation. Unique identifier RFMEFI57814X0073.

References

1. A. Belkov, S. Belov, M. Pavlov, V. Ponomarev, S. Ponomarev and A. Zhukov, Matec Web of Conf. **48**, 02001 (2016)
2. V.N. Zimin, I.M. Koloskov, V.E. Meshkovsky, V.I. Usyukin J. Trans. Mod. and Sim. **30**, 497 (2001), https://www.witpress.com/Secure/elibrary/papers/CMEM01/CMEM01049FU.pdf
3. S. Morterolle, B. Maurin, J-F.Dube, J. Averseng, J. Quirant, J. Aerosp. Sci. and Tech. **42**, 74 (2015)
4. K. Tanizawa, K. Yamamoto, J. Theor. and Appl. Mech. Jap. **53**, 75 (2004)
5. J. Santiago – Prowald, H. Baier, J. CEAS Space, **5**, 89, doi: 10.1007/s12567-013-0048-3 (2013)
6. K. Washizu, *Variational methods in elasticity and plasticity* (Pergamon Press Ltd., Oxford, 1975)
7. B.L. Abramian et al., *Prochnost', ustojchivost', kolebanija. Spravochnik v treh tomah. T.1.* (M: Mashinostroenie, 1968) (in Russian)
8. P. Wriggers, *Nonlinear Finite Element Method* (Springer, Berlin, Heidelberg, 2008)

MATEC Web of Conferences **102**, 01008 (2017) DOI: 10.1051/matecconf/201710201008
V International Forum for Young Scientists "Space Engineering"

Nutation damping of dynamically tuned gyroscopes

Lev Belyanin[1,*], and *Doan Ket* Vu[1]

[1]National Research Tomsk Polytechnic University, 634050 Tomsk, Russia

Abstract. The paper deals with the problem of nutation damping of dynamically tuned gyroscopes (DTG). Unlike the common two-axis gyroscopes, the DTG damping torques caused by dry and viscous frictions are absent in gimbal bearings, while those due to the rotor/gaseous medium interaction are insufficient to provide the required nutation damping. A physical explanation is given to the damping torques induced in the gyroscope by the magnetoelectric torque detectors. Control coils of the latter are included in the output circuits of the electronic amplifiers. The equation is proposed for the calculation of the damping torque. The numerical estimate of the damping coefficient is given in this paper.

1 Introduction

An ideal configuration of a two-axis dynamically tuned gyroscope (DTG) and its ideal tuning allows considering it as a two-axis balanced gyroscope [1, 2].

According to the theory of a rotor two-axis gyroscope, its any forced movement is accompanied by nutation (free oscillations of a gyroscope). In common two-axis gyroscopes, nutation damping in gimbal bearings is mostly provided by the aerodynamic damping torques (interaction between the gyro case and ambient air), dry and viscous friction torques in ball bearings, and dry friction torques in the sliding current collectors and potentiometer angle transducers.

In DTG gimbal bearings the dry and viscous friction torques are practically absent, while damping torques caused by the rotor/gaseous medium interaction are low due to low viscosity and gas pressure inside the internal space of the gyroscope [3, 4]. If the torque detectors of the DTG are not involved in the operation, the damping torques turn to be insufficient to provide the appropriate nutation damping. In this case it is important to estimate the damping torques produced by the DTG torque detectors.

2 Solution of the problem

It is known that the damping torque component is present in a torque produced by any electric motor or torque detector. In order to obtain the equation for the numerical estimate of the damping torque in a DTG produced by the magnetoelectric torque detectors, consider

* Corresponding author: belyaninln@tpu.ru

© The Authors, published by EDP Sciences. This is an open access article distributed under the terms of the Creative Commons Attribution License 4.0 (http://creativecommons.org/licenses/by/4.0/).

their operation at the different orientations of the rotation axis relative to the gyro case having a certain angular rate.

The formula should be derived to determine the value of the torque developed by the torque detectors, the schematic operation of which is presented in Figure 1.

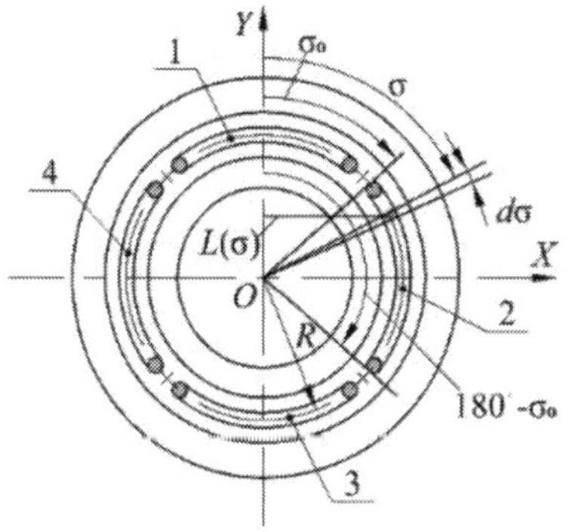

Fig. 1. Schematic operation of the torque detectors.

Figure 1 gives an example of the DTG rotor the circular groove of which allocates four coils or control windings of two magnetoelectric torque detectors. The rotor represents a constant magnet that creates a magnetic field in the circular groove, the lines of forces act along the radius. The circular groove contains the active coil parts mounted to the gyro case by their other parts. During the current flow from the coils to the conductors that form these coils and are placed normally to the rotation axis of the rotor, the axial force action is observed (Ampere's law). The coils *1* and *3* and also *2* and *4* are identical and connected to each other in series such that forces applied to the coils *1* and *3* are always directed oppositely. As a result, the created couple of forces produces the torque around the *OX*-axis. Similarly, the coils *2* and *4* create a couple of forces and the torque around the *OY*-axis.

Let us assume that the magnetic field in the circular groove is described by the inductance *B* the vector of which is directed along the radius, while its value is constant along the whole circle and the groove depth. Each coil has the identical number of turns equaling *w*.

The equation is then obtained for the determination of the torque produced by the torque detector and coils *2* and *4*. The torque is created around the *OY*-axis. Since these coils are completely identical and symmetrical relative to the *OY*-axis, let us discuss only one coil.

The force acting on the segment *dl* of the conductor positioned at σ angle, is determined by

$$dF = I_{tdw} \cdot B \cdot dl = I_{tdw} \cdot B \cdot R \cdot d\sigma, \tag{1}$$

where I_{tdw} is the current in the torque detector winding; *dl* is the length of the segment equaling $R \cdot d\sigma$ in the angular measurement; *R* is the radius of *dl* section at a distance from the rotor axis.

The force *dF* acts on *L*(σ) arm and produces the torque around *OY*-axis. It can be derived from

$$dM'_Y = dF \cdot L(\sigma) = I_{tdw} \cdot B \cdot R^2 \cdot \sin\sigma \cdot d\sigma. \tag{2}$$

The torque produced by the conductor of the coil 2 placed in the circular groove can be obtained from

$$M_Y^{'} = \int_{\sigma_0}^{180^\circ - \sigma_0} I_{tdw} \cdot B \cdot R^2 \cdot \sin\sigma \cdot d\sigma = 2 \cdot I_{tdw} \cdot B \cdot R^2 \cdot \cos\sigma_0. \quad (3)$$

In case the coil contains w conductors and two identical coils contribute to the torque M_Y, we have

$$M_Y = 4 \cdot I_{tdw} \cdot w \cdot B \cdot R^2 \cdot cos\sigma_0 = K_{td} \cdot I_{tdw}, \quad (4)$$

where $K_{td} = 4 \cdot w \cdot B \cdot R^2 \cdot cos\sigma_0$ is the transfer factor of the torque detector.

Let us now determine the damping coefficient provided by the torque detector. For simplicity we ignore the inductance of the detector windings.

Due to the rotation around OY-axis with the angular rate of $d\alpha/dt$, the winding turns of the torque detector cross the lines of forces of the magnetic field created by the constant magnet in the gap. This leads to the electromotive force in the winding turns (the law of electromagnetic induction). The value of the electromotive force de observed in the conductor can be determined from

$$de = \frac{d\Phi}{dt} = \frac{B \cdot dS}{dt} = B \cdot R^2 \cdot \sin\sigma \cdot \frac{d\alpha}{dt} \cdot d\sigma, \quad (5)$$

where $d\Phi$ is the magnetic flux crossed by the conductor with dl length; dS is the area of the magnetic flux crossed by the conductor with dl length during dt time.

The value of the electromotive force observed over the entire length of the winding conductor is given in the following integral formulation:

$$e(t) = \int_{\sigma_0}^{180^\circ - \sigma_0} B \cdot R^2 \cdot \sin\sigma \cdot \frac{d\alpha}{dt} \cdot d\sigma = 2 \cdot B \cdot R^2 \cdot \cos\sigma \frac{d\alpha}{dt}. \quad (6)$$

The electromotive force in the two opposite identical cumulative windings connected in series and containing w turns each, can be obtained from

$$de = 4 \cdot w \cdot B \cdot R^2 \cdot \cos\sigma \frac{d\alpha}{dt} = K_{td} \cdot \frac{d\alpha}{dt}. \quad (7)$$

The electromotive force $e(t)$ acts in a closed-loop system formed by the output circuit of the amplifier, ballast resistance (if any), and ohmic resistance of the control winding. The equivalent circuit of the amplifier and winding is given in Figure 2. The amplifier output stage is assumed to be linear with the output impedance having only an active component R_{ou}. The input is short circuited, thus $e_{ou} = 0$.

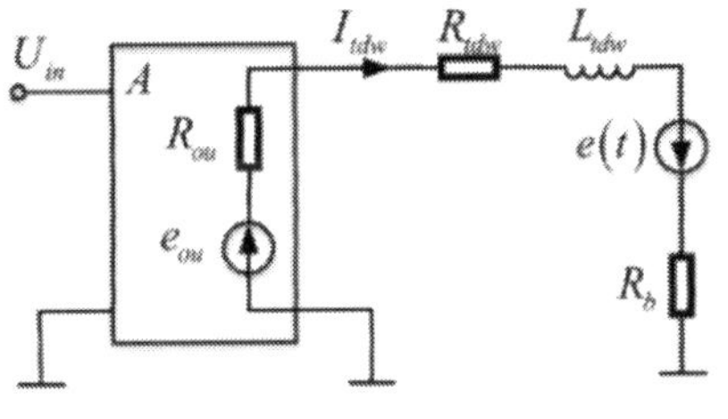

Fig. 2. Equivalent circuit of the amplifier and winding of the torque detector.

Let define the total ohmic resistance as R_Σ, then the current in this circuit caused by the electromotive force *e(t)* can be written in the following from:

$$I_{tdw}(t) = \frac{e(t)}{R_\Sigma} = \frac{K_{td}}{R_\Sigma} \cdot \frac{d\alpha}{dt}. \quad (8)$$

The torque of the detector torque produced by this current is represented as

$$M_{td}(t) = K_{td} \cdot I_{tdw}(t) = \frac{K_{td}^2}{R_\Sigma} \cdot \frac{d\alpha}{dt}. \quad (9)$$

The torque of the detector torque determined by the above written equation is just the desired damping torque. Introducing the damping coefficient provided by the torque detector we have

$$M_{dt} = \mu \cdot \dot{\alpha}, \quad (10)$$

where M_{dt} is the damping torque produced by the torque detector; $\mu = K_{td}^2 / R_\Sigma$ is the damping coefficient produced by the torque detector.

Let us obtain the value of the damping torque produced by the torque detector. For one of the DTG intended for the indication stabilizer systems and produced in the Russian Federation, we have R_{tdw} = 305 Ohm and $K_{td} = 8.72 \cdot 10^{-3}\, N \cdot m \cdot A^{-1}$. Therefore, $\mu = 7.6 \cdot 10^{-8}\, N \cdot m \cdot s$ at R_Σ = 1000 Ohm. In DTG intended for the production of angular rate sensors, the value of the damping coefficient provided by the torque detectors is much higher.

3 Conclusions

The proposed equation allowed the authors to estimate the value of nutation damping produced by the magnetoelectric torque detectors in dynamically tuned gyroscopes. The suggested numerical estimate showed that in a number of cases that damping was insufficient to suppress nutation.

References

1. D.P. Luk'yanov, V.Ya. Raspopov, Yu.V. Filatov, *Principles of theory of gyroscopes* (OAO Elektropribor, St-Petersburg, 2015) (in Russian)
2. G.M. Vinogradov, S.V. Krivosheev, *Dynamically tuned gyroscopes* (KSTU, Kazan, 2008). (in Russian)
3. A. Lawrence, *Modern inertial technology: Navigation, guidance, and control* (Springer-Verlag, New York, 1998)
4. Jeffrey S. Cain, Douglas A. Staley, Glenn R. Heppler, John McPhee, J. Guid. Control Dynam. **29**, 965 (2006)

Application of dual energy method for non-destructive testing of materials designed to work in extreme conditions

Sergei Chakhlov[1,*], *Yanzhao* Wang[1], *Sergei* Osipov[1], and *Victor* Udod[1]

[1]Tomsk Polytechnic University, 634050 Tomsk, Russia

Abstract. The description of the dual energy method (DEM) for non-destructive testing (NDT) of materials and products is presented. It highlights the key factors that determine its accuracy and performance and shows the possibilities for its further improvement. The correlation between the quantum noise level and the DEM precision of the effective atomic number was found.

1 Introduction

The quality control is the most massive operation in the production process, because no detail can be made without measuring its performance. The complexity of the control operations in the industry is increasing dramatically due to the complexity of the new technology and a steady increase requirement of its reliability [1-3]. Thus, in developed countries, the quality control costs are on average from 1 to 3% of the product, and in industries such as defence, nuclear and aerospace, quality control costs increases to 12-18% of [4]. These costs quickly pay for themselves, because due to NDT at all stages of production and acceptance the quality of products radically improving and its reliability is increased.

A variety of practical problems of control, by the target, its content and their terms involves the quality problem solution by various physical methods. In this regard, in the science and practice many NDT methods have developed, which are mainly used for the product examination (bulk defect detection) with the untreated surface and a complex shape. With the development of the nuclear instrumentation, accelerator technology the scope of the technical arsenal of radiation control means greatly enhanced.

Currently the digital radiography occupies one of the dominant positions among the different radiation NDT methods. The term "digital radiography" means the aggregate of radiation NDT methods and diagnostics, which converted at some point the radiation image of the testing object (TO) into a digital signal [5-6]. Subsequently, this digital signal is stored in computer memory and converted into two-dimensional array of measuring data, which can be subjected to various digital signal processing (contrast enhancement, scaling, antialiasing, and the like), and finally, it is reproduced on the display screen as halftone (or colour) image directly perceived by the operator. Currently, digital radiography system

* Corresponding author: chakhlov@tpu.ru

© The Authors, published by EDP Sciences. This is an open access article distributed under the terms of the Creative Commons Attribution License 4.0 (http://creativecommons.org/licenses/by/4.0/).

(DRS) are widely used in industrial and technical flaw diagnosis in medical diagnostics, as well as for screening baggage, carry-on baggage, containers, etc. to ensure the safety of traffic and deter illegal trafficking of prohibited items [7].

Dual-energy method (DEM) is used for improving the efficiency of control in many DRS. Testing object in this method is radiated twice - at two different voltages on the X-ray tube (corresponding to two effective energies of radiation). Radiation detection results are processed by a certain algorithm to obtain an estimate of the atomic number Z (or effective atomic number) material TO [8].

There are various modifications in the implementation of this method, for example - with a single registration TO by two radiation detectors arranged one after the other along the X-rays [9], but instead of high-energy x-ray tubes are used the radiation sources [10].

It is obvious that the possibility of determining the atomic number (or effective atomic number) TO material makes the examination process significantly more informative. This is particularly important during examination of responsible products, designed for operation in extreme conditions, in particular for NDT of structural elements of spacecraft.

2 Basics of dual energy method

In the dual-energy method based on measuring the attenuation of X-rays of two energies, two maximum of some parameters are measured separately, further – DEM parameters. One of the DEM parameter is depended from thickness hand density ρ of TO material, and second one from the effective atomic number. There are several ways to define DEM parameters. The most physically based method associated with the solution of integral equations of two-parameter systems [11]:

$$\begin{aligned} -\ln\frac{\int_0^{E_1} E_A(E)f(E,E_1)\exp(-Ag_1(E)-Bg_2(E))\varepsilon(E)dE}{\int_0^{E_1} E_A(E)f(E,E_1)\varepsilon(E)dE} &= Y(E_1) \\ -\ln\frac{\int_0^{E_2} E_A(E)f(E,E_2)\exp(-Ag_1(E)-Bg_2(E))\varepsilon(E)dE}{\int_0^{E_2} E_A(E)f(E,E_2)\varepsilon(E)dE} &= Y(E_2) \end{aligned} \quad (1)$$

Here $f(E,E_1)$, $f(E,E_2)$ is the energy spectra of X-rays with a maximum energy E_1 and E_2 correspondingly; $g_1(E)$, $g_2(E)$ are the energy dependences of attenuation coefficient of photon radiation for the two physical effects of photon interaction with matter; $E_A(E)$ is an average value of the absorbed energy to a photon with energy E, which is registered by detector; A and B are desired DEM parameters; $\varepsilon(E)$ is the effectiveness of the radiation detecting energy E for the detector; $Y(E_1)$, $Y(E_2)$ are first and second informative parameters (converted radiation measurement results corresponding to both maximum energies E_1 and E_2).

Prevailing effects of the interaction of photon radiation with matter in the range of X-ray energies up to 200 keV is the photoelectric effect and the Compton Effect (incoherent scattering). Therefore according to [11-12] we have:

$$A = Z^{3.5}\rho h; \quad (2)$$

$$B = \rho h \quad (3)$$

$$g_1(E) = C_1 E^{-2.8},\ E \geq 0.02\ \text{MeV} \quad (4)$$

$g_1(E)$ is the energy dependence of the photoelectric effect; C_1 is a constant;

$$g_2(E)=C_2\left\{\frac{1+\alpha}{\alpha}\left[\frac{2(1+\alpha)}{1+2\alpha}-\frac{ln(1+2\alpha)}{\alpha}\right]+\frac{ln(1+2\alpha)}{2\alpha}-\frac{1+3\alpha}{(1+2\alpha)^2}\right\} \quad (5)$$

$g_2(E)$ is the energy dependence of the Compton effect, where $\alpha=E/0.511$, E is energy in MeV; C_2 is a constant.

Kramers formula as well as its various modifications used traditionally to describe the X-ray energy spectrum in the region of maximum energy of 200 keV. Equations (1) - (5) are the basic analytical relationships that express the essence of DEM using sources with a continuous energy spectrum of radiation with maximum energy in the spectrum up to 200 keV.

3 Factors determining the accuracy and quick dual energy method

From (1) - (5) it follows directly that DEM accuracy is mainly determined by the following factors:

- the accuracy of approximation the interaction of photon radiation with matter in the energy range of X-rays up to 200 keV photoelectric effect and the Compton effect;
- accuracy of the measurement of informative parameters $Y(E_1)$, $Y(E_2)$, is determined, in particular, the accuracy (capacity) of analog-to-digital conversion of output processes of the radiation detector and the statistical fluctuations of the radiation detection results caused by the quantum nature of radiation;- accuracy of estimation of maximum energy in the spectrum of the radiation source;
- setting up of the energy spectrum of the radiation generated by the source;
- accuracy of estimation of the average value of the energy absorbed by the detector;
- accuracy of estimation (calculation) radiation detection efficiency of the detector;
- precision of parametric solution of the integral equations.

Examination of each of these factors, individually and in various combinations, represents a significant scientific problem. In particular, in the field of radiological methods the efforts of many experts focused on the search for more accurate analysis of the cross section of the photoelectric energy. For example, instead of formula (4) in [13] there was proposed a formula like $g_1(E)=C_1E^{-3}$. It should be noted that the preference in the choice of a particular analytical relationship is usually set empirically for a particular group of materials to be recognized.

It is quite obvious that the DEM accuracy can be somewhat improved if the formula (1) to add one more, taking into account the effect - coherent scattering. In particular, based on the data presented in [14], the energy dependence of the coherent scattering type analytically approximated function adequately $f_{coh}(E)=1/E^2$.

The solution of the integral system of parametric equations (1) is one of the biggest difficulties of DEM. In [11] provides a method for the solution of this system based on the concept of a two-dimensional line-level functions, and offered its simple physical implementation.

With regard to DEM performance, then it will be determined by the following steps of determining the parameters of the algorithm A and B. In the first stage, calculated value functions describing the energy dependence of the cross sections of the photoelectric effect and incoherent scattering. In the second stage there is calculated a value of the function approximating the actual energy spectrum of the radiation source. Then function values are calculated, describing the detection efficiency of the radiation detector, and estimated a value of the function that describes the average value of the energy absorbed by the detector. Then the integrals are evaluated in the system of integrated-parametric equations

(1). Finally, the parameters *A* and *B* are calculated. Each of these procedures are characterized by a number of arithmetic operations that determine the actual speed of data processing algorithm based on the DEM. It should be noted that the total number of arithmetic operations significantly depends on the method of numerical integration and convergence speed.

4 Conclusions

At the present stage of the development of digital radiography systems, implementing DEM, there were analysed the images from 256 × 256 to 1024 × 1024 elements, and more, that is, in the pre-processing stage is necessary to find solutions from 65536 to 1048576 or more systems integrated-parametric equations of type (1). Thus, the total number of computational operations required for processing of all data set (dual digital radiographs), can be very large. Therefore, the entire process will not take place in "real time" and, consequently, to minimize the number of arithmetic operations at each stage of the overall algorithm is a very important task.

Thus, further DEM improvement can be done in many different ways. In particular, it is important to establish the quantitative dependence of quantum noise on the accuracy of the material's atomic number estimation.

The correlation between the quantum noise level and the DEM error of the effective atomic number for the aluminium plates, used for the aircrafts sheathing, was found as a result of our theoretical numerical studies.

References

1. A.I. Mamaev, V.N. Borikov, V.A. Mamaeva, T.I. Dorofeeva, Protection of Metals, **41(3)**, 254 (2005)
2. S.V. Muravyov, V.N. Borikov, S.A. Kaysanov, *18th IMEKO World Congress*, **1**, 77 (2006)
3. E.V. Golodnykh, V.N. Borikov, *SIBCON*, 6693635 (2013)
4. V.V.Kluev, *Pribory dlja nerazrushajutshego kontrolja materialov i izdeliy* (Mashinostroenie, 1986) (in Russian)
5. Doucette Ed., Mater. Eval **63**, 10 (2005)
6. P. Charnock, P.A. Connolly, D. Hughes, B.M. Moores, Radiat. Prot. Dosim, **114**, 1 (2005)
7. V.A. Udod, Ya. Van, S.P. Osipov, et al, Rus. J. of Nondestr. Testing **52**, 9 (2016)
8. J.S. Park, J.K. Kim J., Korean Phys. Soc. **59**, 4 (2011)
9. J.Ch. Han, H.K. Kim, D.W. Kim, S. Yun, H. Youn, S. Kam, J. Tanguay, I.A., Cunningham Current Applied Phys. **14**, 12 (2014)
10. Y. Gil, Y. Oh, M. Cho, W., Namkung Applied Radiation & Isotopes **69**, 2 (2011)
11. S.P. Osipov, E.E. Libin, S.V. Chakhlov, O.S. Osipov, A.M. Shtein, NDT & E Int. **76** (2015)
12. R. Alvarez, A. Macovski, Proc. SPIE **96** (1976)
13. Zh. Ying, R. Naidu, C.R. Crawford, J. of X-Ray Sci. & Tech. **14** (2006)
14. O.F. Nemec, Yu.V., Gofman, *Spravochnik po jadernoi fizike* (Kiev, 1975) (in Russian)

MATEC Web of Conferences **102**, 01010 (2017) DOI: 10.1051/matecconf/201710201010
V International Forum for Young Scientists "Space Engineering"

Implementing CDIO Approach in preparing engineers for Space Industry

Yury Daneykin[1,*], *Natalia* Daneikina[1], *Michail* Solovyov[1] and *Arun* Patil[2]

[1]Tomsk Polytechnic University, 634050 Tomsk, Russia
[2]Deakin University, Australia

Abstract. The necessity to train highly qualified specialists leads to the development of the trajectory that can allow training specialists for the space industry. Several steps have been undertaken to reach this purpose. First, the University founded the Space Instrument Design Center that promotes a wide range of initiatives in the sphere of educating specialists, retraining specialists, carrying out research and collaborating with profiled enterprises. The University introduced Elite Engineering Education system to attract talented specialist and help them to follow individual trajectory to train unique specialist. The paper discusses the targets necessary for achievement to train specialists. Moreover, the paper presents the compliance of the attempts with the CDIO Approach, which is widely used in leading universities to improve engineering programs.

1 Introduction

Tomsk Polytechnic University aims at training highly qualified specialists in various technical areas. One of the quite modern area is space industry that needs professional solutions and experts. Therefore, there is an acute necessity to find ways to improve the quality of educational programs and training of specialists.

In 2011 Tomsk Polytechnic University joined CDIO Initiative - the major international project on the reform of engineering education, which began in October 2000 at the Massachusetts Institute of Technology (MIT, USA) with the participation of scientists, teachers and industry representatives [1]. The intention was to improve the quality of education constantly using all the tools possible. The CDIO approach to engineering programs modernization provides the innovative framework for producing the new generation of engineers. Implementation of the CDIO approach in practice requires significant changes in program development, realization and assessment. To design and implement new programs the university faculty have to improve their competences in curriculum development in compliance with the CDIO Syllabus, in integrated learning experiences, in active and experiential learning, and assessing student learning [2]. Authors believe that implementing CDIO Approach will highly increase the quality of education. Particularly, the improvements are necessary in aerospace industry, as the technologies used in this industry have been developing rapidly and specialists with particular training are on demand.

* Corresponding author: daneykin@tpu.ru

© The Authors, published by EDP Sciences. This is an open access article distributed under the terms of the Creative Commons Attribution License 4.0 (http://creativecommons.org/licenses/by/4.0/).

2 Training of specialists in space industry

Tomsk Polytechnic University has Space Instrument Design Center whose aim is contributing to the improvement of specialists' training and research in the sphere of space industry. It states the following purposes [3]:

- Advanced training of world-class experts in the field of Space Device Engineering (masters and PhDs).
- Carrying out research, development work in the field of design, engineering, production and testing of space technology and ground-based equipment in tight connection with the enterprises.
- Implementation of the results of joint innovative research and educational activities
- Development of various forms of post-graduate education, including vocational training and skills development in the field of new space technology and ground-based equipment

In addition, the Center complete the following tasks:

- Providing targeted training for enterprises on the agreed individual educational trajectory.
- Preparation of training and methodological support for implementation of the Master's programs developed by request and with the participation of enterprises.
- Organization of educational and production practices for students that contribute to the maximum approximation to the real conditions of production and research activities
- The formation and consolidation of practical skills of future specialists and practical skills obtained through theoretical training, including by working in multidisciplinary teams.
- Carrying out research activities in the interest of enterprises.
- Promote the establishment of joint enterprises and innovative structures, the establishment of joint educational and research laboratories.
- Implementation of joint cross-disciplinary research and development work and innovative projects with the involvement of the bachelors, masters and PhD students.

3 Targets to achieve

Aerospace is the industry of cutting-edge technologies. The highly integrated yet inter-disciplinary nature of aerospace manufacturing requires graduates with a broad knowledge of technologies, professional skills, and a system-level mind set [4] [5]. The purpose of the authority of the university is the modernization of the basic engineering education - provision of training graduates for complex engineering activity.

The faculty states the following targets to achieve a high level of training.

1. Preparedness to complex engineering activity, which includes the readiness of enterprises to take part in the organization of students' project work (themes of the projects, expendable materials, supervisors) and efficiency of the realization of project activities in the departments.

2. Practical part of training. This includes the readiness of enterprises to take students for traineeship (providing them with actual themes, supervision), terms and practical content of traineeship, efficiency of traineeship organization.

3. Elite Engineering Education (EEE) system provides the new structure of program. For the last year, the number of the students involved increased that confirms the popularity of the format. The formation of education track ("Innovations", "Research", and "Technology") is individual that also attracts the potential participants.

4. Set of measures on interaction with enterprises, which includes the readiness of enterprises to participate in the life of university. It will influence the efficiency of realization of this work in university (monitoring, taking management decisions).

5. Targeted training means the readiness of enterprises to sign contracts with senior students on targeted education, possible social support, organization and monitoring of this work in university.

6. Development of universal competences demanded by employers, taking into account the opinion of enterprises about the universal competences (Federal State Education Standard, additional competences), necessity to know English.

The main objective is to develop students' interest to engineering, strengthen motivation to learning in accordance with the selected direction or specialty, as well as to provide the basis for the development of practical skills.

4 Elite engineering education system

Elite Engineering Education system was introduced at TPU in 2004 on the basis of specialists' and undergraduates' educational programs. In 2014 it was first introduced for Master's degree students. The purpose of EEE system is training of the future leaders of the engineering profession for innovation and entrepreneurship. EEE program is a program of additional training. Learning outcomes include profound knowledge of fundamental sciences (Mathematics, Physics, Economics), competences in the sphere of engineering entrepreneurship, leadership qualities and teamwork skills, holistic worldwide vision and formation of axiological complex thinking.

5 Interaction with enterprises

As we mentioned above close interaction with enterprises. They are involved in a number of activities helping to understand the demands of the market, need in technologies, direction for the improvement of educational programs, ways to organize learning environment and practical training, assess the competitiveness of educational program, the set of competences necessary for the graduates. So areas of interaction with enterprises include as follows.

Design of General educational program (GEP)

- Participation in the design of content of educational modules
- Evaluation of resources for realization of educational program (equipment, site, software etc.)

Realization of GEP

- Open lectures with participation of representatives of enterprises
- Representation of professional modules of GEP (formation of professional and universal competences), e.g. for acquiring blue-collar jobs
- Discussion of the themes of final assignments
- Organization of training experience
- Supervision of final assignments
- Traineeship of teachers

Assessment of the quality of GEP

- Assessment of students' competences (traineeship, course assignments, final assignments)
- Assessment of final assignments (participation in state examination commission)

The improved educational programs provide the achievement of all these competences. Moreover, TPU is now working to rethink educational programs based on CDIO Syllabus, which will enable the creation of a modern level of educational standards

6 Compliance with CDIO Standards

Let us look at the compliance of the targeted educational program with the CDIO Standards. Mostly we were concerned with the Standards 2, 3 and 8.

Standard 2 underlines the importance of engineering reasoning and problem solving, experimentation and knowledge discovery, system thinking, creative thinking, critical thinking, and professional ethics. Interpersonal learning outcomes include teamwork, leadership, communication, and communication in foreign languages, conceiving, designing in enterprise etc. Standard 3 says about reaching the targeted learning outcomes in disciplinary knowledge and skills, where faculty play an active role in designing the integrated curriculum by suggesting appropriate disciplinary linkages, as well as opportunities to address specific skills in their respective teaching areas. Therefore, we successfully used content-based instruction, learning professional skills through tight work with enterprises and the system of elite educational training with individual trajectory of study. Standard 8 suggests using methods of active learning, which were applied when realizing the real team projects, which were developed both in laboratories and in enterprises after discussion with experts.

7 Conclusions

Space industry needs specialists with both broad and deep knowledge, professional competences and personal skills. Special Space Instrument Design Center founded in TPU sets a number of tasks that highly contribute to the development of science for further implementation in the industry, including doing research by the order of enterprises.

Special Elite Engineering Education system contributes to the formation of a specialist with specific skills, developed with the view of future work.

In addition, the authors believe that CDIO Approach, which has been widely used recently to improve and develop engineering programs, can contribute to the improvement of Engineering programs aimed at the training of highly-qualified specialists in the area. The comprehensiveness of Approach make it a good tool that can be implemented in engineering programs in various aspects.

References

1. E. Crawly, J. Malmqvist, *Rethinking Engineering Education* (Springer Science and Business Media, 2007)
2. A. Kriushova, et al, *INTED Proc.* (2016)
3. *TPU official site*, URL http://portal.tpu.ru/departments/centre/noc_kp
4. Ir. Aldert Kamp, *Proc. of the 8th Int. CDIO Conf.* (2012)
5. Yu. Daneykin, E. Serebryakova, M. Solovjev, I. Abrashkina, Nauka i obrazovanie, **7**, 336 (2016)

MATEC Web of Conferences **102**, 01011 (2017) DOI: 10.1051/matecconf/201710201011
V International Forum for Young Scientists "Space Engineering"

Research of the launch vehicle design made of composite materials under the aerodynamic, thermal and acoustic loadings

Denis Davydovich[1], *Michael* Dron[1], *Konstantin* Zharikov[1], and *Yulia* Iordan[1,*]

[1]Omsk State Technical University, 644050 Omsk, Russia

Abstract. The experimental research of the carbon composite material sample of payload fairing half structural element was carried out under different types of loading. Mathematical and physical modeling of the sample loading using aerodynamic flow was conducted. Heat loading was researched by the method of a thermal analysis during which typical heat dots corresponding to the changes in the sample structure were determined. Ultrasonic influence on the sample characteristics was considered. As a result, the value of heat leak to the structure surface while moving in the atmospheric phase of the descent was determined.

1 Introduction

The technique described in [1, 2] is proposed for solving the problem of minimizing the exclusion zones under the exhaust parts of the rockets, made of composite materials (CM). According to this technique, we suggest to modify the material of units made of CM, such as the payload fairing half, due to additional energy effects for achieving the fine phase of the structure or for completing the combustion. The method is based on determining the missing quantity of heat as the difference between the incoming and desired heat fluxes. Pyrotechnic composition (PC) as possible additional sources are considered. The assessment of the required mass of the PC and the temperature increase of the structure are presented in [3]. This technique is described in [4], where titanium parts are used as the destructive elements.

The results given in [5, 6] show that ultrasonic treatment leads to a change in the structure of a solid, and can significantly reduce energy consumption for the implementation phase transitions.

The aim of this work is to study the influence of additional energy sources on the structure of the sample with the purpose to achieve maximum destruction.

* Corresponding author: iordanyuliya@gmail.com

© The Authors, published by EDP Sciences. This is an open access article distributed under the terms of the Creative Commons Attribution License 4.0 (http://creativecommons.org/licenses/by/4.0/).

2 Materials and methods

The sample of the payload fairing half structural element is a carbon fiber based on carbon tape LU-P-0,1/A and binders. Table 1 presents the composition of the binder according to [7].

Table 1. The composition of the binder.

Component name	Mass fraction, %
Resin EN-6	77.0
Furfurylglycidyl ether	15.0
The catalyst UP 605/3	3.0
Resin SF-341-A	5.0

Mathematical modelling of the aerodynamic loading was carried out according to [8-10] for the cases of flight of payload fairing half at angles of attack of 0^0, 90^0 and 270^0. The motion of the centre of mass, without regard to motion relative to the centre of mass, without taking into account the rotation of the Earth is missing, was considered.

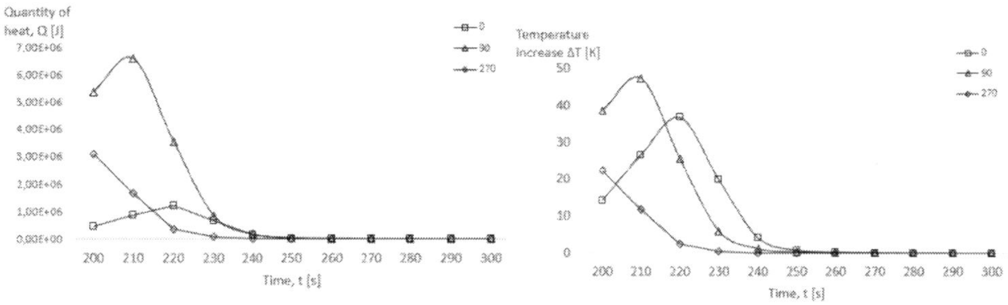

a b

Fig. 1. a) Graph of the quantity of heat received by the payload fairing half during the descent from time; b) growth graph of the surface temperature of the structure during the descent.

For physical modelling of aerodynamic heating a bench, providing pre-heating of the sample to a temperature corresponding to a given height was designed. This bench can make the required ram-air flow to simulate the process of blowing with the purpose to determine the magnitude of heat losses. The sample with sizes 40x40x3 mm was pre-heated by 0.8 mm-thick nichrome thread with to a temperature of 3000 C. Thermocouples, fixed in front of the sample and behind it, transferred the data to a personal computer with an interval of 0.5 s. The system of flow simulation provided a given rate of air leak. The experiments were conducted at atmospheric pressure and ambient temperature of 22.5^0C.

The study of thermal loading of the sample was carried out by using a thermal analysis instrument STA 449 F3 Jupiter (Netzsch, Gremany). The sample with mass 19.595 mg was placed into an open corundum crucible and was heated with a heating rate of 10 K/min in airflow 100 ml/min.

The experimental bench described in [11], allows to simultaneously simulating the aerodynamic, thermal and acoustic impacts on the test sample. The supply of extra energy in the form of ultrasound exposure leads to a change in the coefficient of thermal conductivity, burning rate, etc., which contributes to the achievement of the research goal.

 DOI: 10.1051/matecconf/201710201011

3 Results

The results of mathematical and physical modelling are represented in Figures 1 and 2, relatively. During thermal decomposition of the sample on the curve thermogravimetry (TG) three stages of mass loss are determined; the second and third stages of the mass loss curve of differential scanning calorimetry (DSC) correspond to successive exothermic peaks (Figure 3). Based on the results of the preliminary experimental studies, ultrasonic treatment on the sample leads to the increase of heat transfer coefficients from the sample to the flow up to 15% of the nominal value.

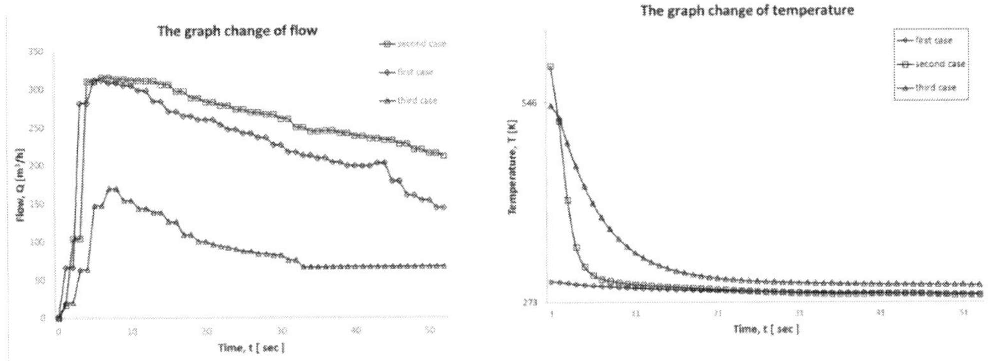

a) b)

Fig. 2. a) Graph of consumption of the aerodynamic flow; b) graph of temperature change of the sample surface during the blowing of the aerodynamic flow.

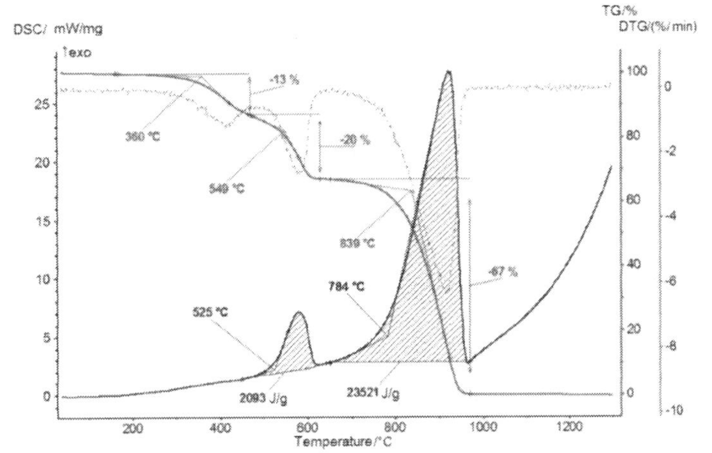

Fig. 3. The thermogram of the mass change (TG), rate of mass loss (DTG) and the rate of heat emission (DSK).

4 Discussion

Based on the calculation results presented in Figure 1, it is shown that the value of heat leakage to the surface of the structural element of the payload fairing half is changes depending on the angle of attack and the rate of descent. The maximum increase in temperature from aerodynamic heating in the part from 60 to 0 km for the considered trajectory case is 47 degrees (without entrainment of heat).

The experimental studies on modelling of aerodynamic load have shown that there is a big heat loss from the surface of the sample, particularly, from the heating element.

As it can be seen from the experiments on thermal loading, while increasing temperature up to 1000^0C, coal threads constituting the carbon tape of the test sample design, are completely burned. This process is accompanied by considerable heat emission, the value of which is comparable to the calorific value of coal.

The mechanism of changes in the structure of the sample under the acoustic effect is known and is explained by the process of excitation and propagation in layers of the ultrasonic vibrations material, which are periodically alternating phases of compression and rarefaction of the domain. A possible subject of the further research is the determining of the optimal parameters of acoustic influence leading to the destruction of the structure.

5 Conclusions

According to the study, we can draw the following conclusions:
1. Simulation of aerodynamic flow is accompanied by significant heat losses that must be considered when choosing sources of thermal loading.
2. To bring the structure to finely disperse phase is required to heat the material of the structure to a temperature of 1000^0C.
3. Ultrasonic effect contributes to the destruction of the sample at lower energy cosumption. To bring the sample to the maximum possible destruction we need to find the certain parameters of ultrasonic influence.

It is necessary to define the optimal ratio of acoustic and thermal energy acting on the sample structure to provide its destruction.

The authors express their acknowledgment to the Professor V. Trushlyakov

The research was supported by the RSF grant for the project "Development of scientific-technical fundamentals of the combustion of the separated elements of space rockets with the goal of reducing the acreage of the impact areas" the Agreement No. 16-19-10091 18.05.2016.

References

1. V. Trushlyakov, Ya. Shatrov, D. Lempert, Yu. Iordan, V. Zarko, RF Patent № 2581636 (2016)
2. V. Trushlyakov, Ya. Shatrov, , RF Patent № 2585395 (2016)
3. D. Lempert, D. B., Trushlyakov, V. I. and Zarko V. E., CESW **51**, 5, 619 (2015)
4. K. Monogarov, A. Pivkina, N. Murav'ov, *Combustion and explosion* 7, 327 (2014)
5. H. Puga, S. Costa, J. Barbosa, S. Ribeiro, M. Prokic, J MATER PROCESS TECH **211**, 11, 1729 (2011)
6. V. Trushlyakov, V. Kudentsov, I. Lesnyak and etc., *Project Report* № 2.1.2/13519, OmSTU (2011)
7. *State Standard* 28006-88 (1989)
8. A. Tewari, J Spacecr Rockets **46**, 2, 299 (2009)
9. A. Glazunov, Tomsk State University Journals **4**, 79 (2011)
10. D. Balakrishnan, J. Kurian, J Spacecr Rockets **51**, 4,1319 (2014)
11. V. Trushlyakov, K. Zharikov, M. Dron', Yu. Iordan, D. Davydovich, , RF Patent № 2016134556 (2016)

MATEC Web of Conferences **102**, 01012 (2017) DOI: 10.1051/matecconf/201710201012
V International Forum for Young Scientists "Space Engineering"

Additive effects under the series of EOS in space application VLSI circuits

Nikolai Diatlov[1,*], *Petr* Skorobogatov[1], and *Konstantin* Epifantsev[1]

[1]National Research Nuclear University MEPhI, 115409 Moscow, Russia

Abstract. One of the problems of space technology is the spacecraft on orbit charging effect. Series of EOS (electrical overstress) are caused by internal charging affect VLSI (very large-scale integrated) circuits, which may lead to its damage. The results of the EOS series with energy below the threshold of failure for modern submicron VLSI circuits design are presented. The obtained results confirm the possibility of accumulation of the effects of damage from exposure of EOS series in modern VLSI circuits and allow you to get the dependence describing the additive nature of damage the VLSI circuits during exposure to subthreshold EOS. The obtained dependence agrees well with the Arrhenius equation, which indicates the thermal nature of the damage under the series of subthreshold energy EOS. The method of the VLSI circuits testing is proposed to determine the level of the VLSI circuits EOS hardness to the effects of multiple different pulsing voltages.

1 Introduction

The space industry has currently faced the problem of the spacecraft lifetime. One reason for the decrease in lifetime of the spacecraft is the charging effect [1].

The charging effect of the spacecraft is caused by exposure to a flow of charged particles near-Earth plasma, the solar wind, galactic radiation and light. Different parts of the spacecraft accumulate charge unevenly as a result of its non-uniform irradiation and the use of different materials in the structure. Over time, it has formed between the parts defined potential difference, and electrostatic forces act. When the electric field strength exceeds the critical (for near-Earth space plasma $E_{max} \sim 10^7\ V/m$) electrical discharge occurs between the parts of the spacecraft, which can lead to malfunctions and damages of sensitive electronics of the spacecraft.

The internal charging effect is caused by the accumulation of charge by particles penetrating within the spacecraft body. Charged particle energy over 100 keV penetrate the spacecraft casing and is absorbed by dielectrics and ungrounded conductors located in a part of the sensitive electronics (Figure 1). Despite the relatively small flow of high-energy charged particles and small amplitude of discharge pulse internal charging effect can have a significant impact on the operation of the spacecraft [1]. The reason is the direct impact of EOS on sensitive components [2].

* Corresponding author: nsdyat@spels.ru

© The Authors, published by EDP Sciences. This is an open access article distributed under the terms of the Creative Commons Attribution License 4.0 (http://creativecommons.org/licenses/by/4.0/).

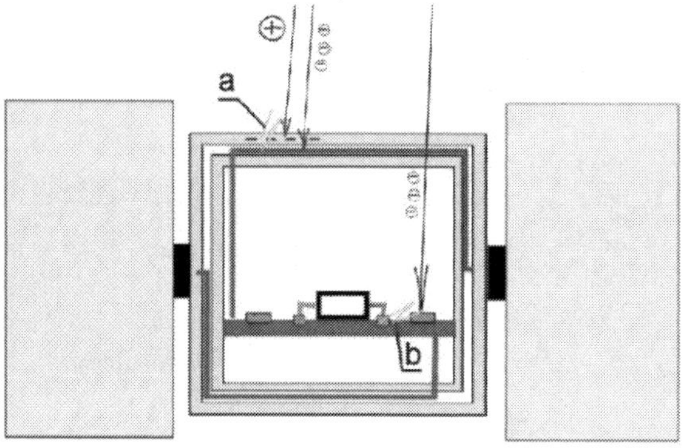

Fig. 1. EOS in outer shell of the spacecraft caused by the external charging effect (a) and EOS between terminal of VLSI circuit and ungrounded conductor caused by the internal charging effect.

Electrical discharges caused by internal charging effect leads to the appearance of the voltage pulses at terminals of electrical components. Amplitude of the voltage pulses may be up to ten and hundreds volts [2].

During the entire period of operation of the spacecraft many internal electrical discharges occur. Over 10 years of operation, their number can reach about 10^4 [3]. Thus, the series of electrical discharges caused by the internal charge effect will affect terminal of a single VLSI. Under the influence of the additive effect of damage accumulation in VLSI circuits, failure may occur when the EOS energy is below the threshold of failure [4].

The article presents the results of an experiment on the effects of a series of subthreshold energy pulse voltage CMOS IC CD4007. This data confirms the presence of additive effects in CMOS circuits [5].

For a more detailed research of this effect, the impact of EOS series on the behaviour of two types of modern controllers was studied.

2 Materials and methods

The specialized pulse voltage generator (EMI-0501) was used to estimate the EOS behavior. The EOS pulses were applied between the ground and appropriate pin. The impact of EOS was repeated until the failure of VLSI circuit was recorded.

Test procedure and generator parameters are described in [6].The LVSI circuit under test was connected through the buffer unit to a storage oscilloscope and a functional control module.

3 Results

Dependence $N(U_{EOS})$ was obtained for each type of microcontrollers, where N - number of voltage pulses of amplitude effects U_{EOS} until sample failure. These dependences are shown in Figure 2.

The following function (1) provides the best fit to obtained data for all types of microcontrollers.

$$N(U_{EOS}) = \exp(b(\frac{1}{{U_{EOS}}^2} - \frac{1}{{U_O}^2})) \quad (1)$$

Where U_O - amplitude threshold of the failure for this type of LVSI circuit, b - parameter of approximation, depending on the product type.

Linear approximation (2) in semi-logarithmic scale on the x axis was obtained based on the results (Figure 3).

$$N(x = b(\frac{1}{U_{EOS}^2} - \frac{1}{U_O^2}) \qquad (2)$$

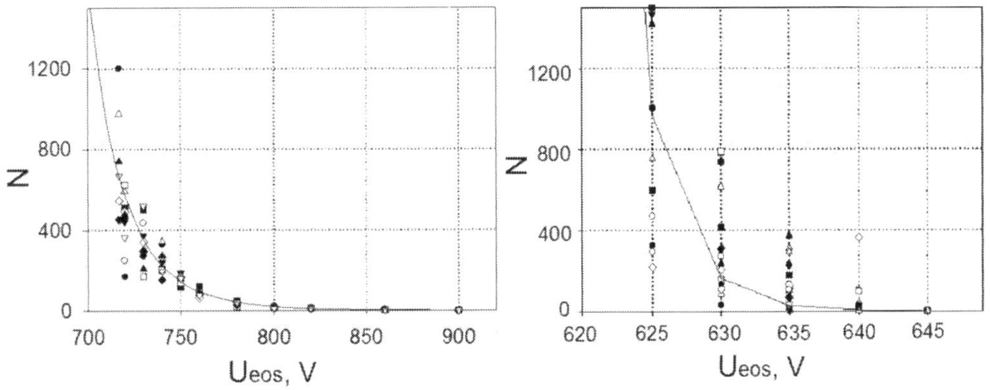

Fig. 2. Dependences $N\ (U_{EOS})$ for MC ATtiny13A (left) and MC PIC24F16KA101-I / MQ (right).

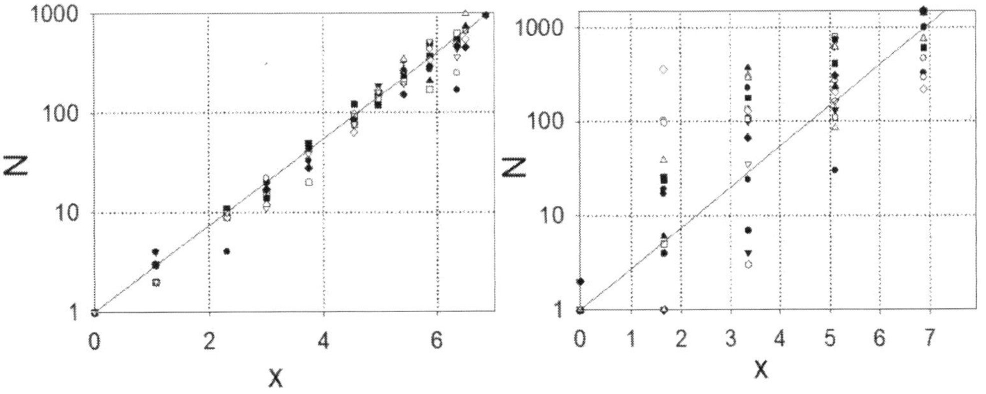

Fig. 3. Linear approximation $N(x)$ for MC ATtiny13A (left) and MC PIC24F16KA101-I / MQ (right).

4 Discussion

The dependence shown in Figure 2 shows that for different types of VLSI circuits investigated voltage range (for which the number of the voltage pulse impact was relatively small $N(U_{EOS}) < 1000$) varied considerably. From this, it follows that for VLSI circuits experiencing during operation a large number of EOS effects, critical voltage amplitude U_{CR} may be substantially below the threshold of failure when exposed to a single EOS pulse.

Therefore, the design of electronics is necessary to consider EOS caused by internal charging, and their impact on the performance of the element base, especially in the long term of operation.

The results of approximation (1) comply well with the Arrhenius equation. (3)

$$v^{-1} = \exp(\frac{E_a}{kT}) \qquad (3)$$

v^{-1} - the inverse of the speed energy-dependent processes in the semiconductor and corresponds to the number of impacts N and the kT - corresponds to the stress caused by the EOS pulses (U_{EOS}^2 determines the power that is in the chip element). The difference in the reverse voltages of squares obtained dependence is related to the ratio of the electrical stresses at the points U_O and U_{EOS}.

5 Conclusions

From experimental results it follows that it is desirable to conduct tests on the sensitivity of the components that make up the spacecraft electronics, to the effects of EOS series.

For dependence $N(U_{EOS})$ there is no need to destroy a large number of chips. It is sufficient to determine the amplitude threshold failure U_O and obtain the value $N(U_{EOS})$ at one point. This makes it possible to carry out tests on the IC (integrated circuits) sensitivity to EOS series of subthreshold energy at low cost.

The obtained dependence complies well with the Arrhenius equation, which indicates the thermal nature of the damage when exposed to series of subthreshold EOS.

References

1. D.L. Mulville, *NASA Technical Handbook,* **4002** (1999)
2. A.R. Frederickson, *Spacecraft Charging Technology, Proceedings of the Seventh International Conference held 23-27 April*, 417 (2001)
3. W. Kim, I. Jun, M. Kokorowski, IEEE Trans. on Nucl. Sci. **57**, 3143 (2010)
4. M. Gorlov, A. Stroganov, A. Adamian, ChipNews **1**, 34 (2001)
5. P. Skorobogatov, K. Epifantsev, *Conf. Scientific session of the MEPhI, theses*, 108 (2013)
6. P.K. Skorobogatov, *Proceedings 4th European Conference on Radiations and Its Effects on Components and Systems (RADECS 97)*, 174 (1997)

The principle of gravity-inertial orientation

Victor Dmitriev[1], and *Ruslan* Frolov[1,*]

[1]National Research Tomsk Polytechnic University, 634050 Tomsk, Russia

Abstract. The important problem of gyrocompass without drift creation is considered in this article. The designing device allows to define the accurate information about mobile and stationary objects orientation in the case of long-term (months, years) absence of the determining the cardinal direction possibility. This article focuses on the idea of gyrocompass without drift creation, which works on the principle of gravity-inertial orientation at the theoretical level.

1 Introduction

Ground orientation is the position indication of man and other different objects on the ground relative to the cardinal directions and landmarks. The science of navigation considers these issues. In modern science and technology, interdisciplinary ideas have the greatest development.

2 Formulation of the problem

Nowadays, the theory of inertial navigation systems emerged as the independent branch of applied mechanics with its objectives and methods. To have a gyrocompass with minimal drift, and ideally - without drift is the principal technical problem, which has always been and is currently exist in the field of navigation systems [1-4].

3 Technical research

Gyroscopic action of the rapidly spinning bodies consists in their capacity to resist to the external forces, changing their position in the inertial space [5-7].

The apparent anomaly in motion of the gyroscope is explained by the appearance of Coriolis (complementary) acceleration.

Coriolis acceleration occurs when the relative and translational motion are simultaneously applied to the object (point), and consequently Coriolis force acts on the object, located on the Earth surface:

$$F_{cor}=m \cdot \dot{\upsilon}_C=2 \cdot m \cdot \ddot{\upsilon}_C \cdot \Omega_E \cdot cos\, \varphi, \quad (1)$$

* Corresponding author: phrolov1994@mail.ru

© The Authors, published by EDP Sciences. This is an open access article distributed under the terms of the Creative Commons Attribution License 4.0 (http://creativecommons.org/licenses/by/4.0/).

where, m – mass of the object; Ω_E – the angle rate of Earth's rotation; $\dot{\upsilon}_C$ – Coriolis (complementary) acceleration.

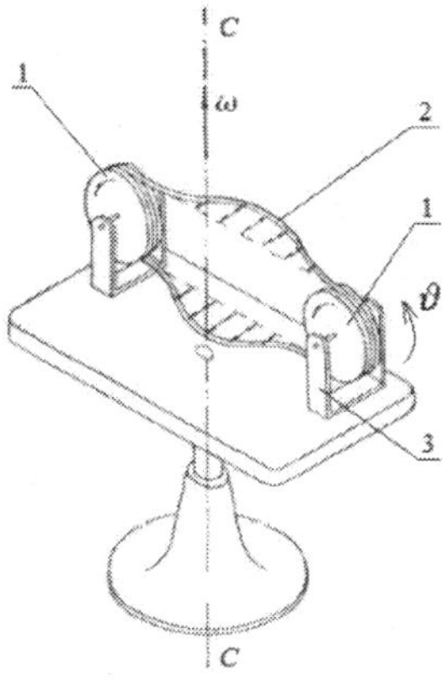

Fig. 1. Installation for Coriolis acceleration demonstration.

Figure 1 shows the main parts of installation, they are 1 – pulley, 2 – belt, 3 – table. Pulleys 1 start to rotate with fixed table. In this case, the straight sections of the belt will move along table with some relative rate. If the table is forced to rotate around the C-C axis with the rate ω, interaction of the inertial and elastic forces of the belt will provide the parabolic character of the belt deformation. To be more convincing, that it is not the action of the centrifugal forces, but it is the action of the Coriolis acceleration, C-C axis does not coincide with the axis of the table rotation [8].

A similar pattern is observed, if the undulatory motion of the object with weight m along the place vertical is made. In this case, the object acquires the linear velocity ϑ_r along $\bar{g}$. The value of linear velocity in time is represented by the expression:

$$\upsilon_r = \omega \cdot z_m \cdot \cos \omega t, \tag{2}$$

where, z_m – amplitude of swell; ω – pulsatance; t – time.

Then (Figure 2), there are two motions: the rotational (translational) from the Earth's rotation and the linear (relative) which acts along the place vertical. Consequently, the compass force F_{com} appears [9], which can be written as:

$$F_{com} = 2 \cdot m \cdot \Omega_E \cdot \cos \phi \cdot \omega \cdot z_m \cdot \cos \omega t, \tag{3}$$

where, m – mass of the sensitive element; Ω_E – the angle rate of Earth's rotation; φ – site latitude; z_m – amplitude of swell; ω – pulsatance of the periodic motion of the sensitive element.

Consider the general case of the point (sensitive element) rectilinear motion in the mechanical system, which has three degrees of freedom (Figure 2).

Applying the Lagrange method, we will have the differential equations, which describe forced movement of the sensitive element in vertical direction. Kinetic energy T, potential energy V, scattering function F, generalized forces (Q_P, Q_R, Q_F) are determined for this mechanical system as:

$$T = \frac{1}{2} m \cdot \dot{q}^2, \qquad Q_P = -\frac{\partial P}{\partial q} = -cq,$$
$$V = \frac{1}{2} c \cdot q, \qquad Q_R = -\frac{\partial F}{\partial \dot{q}} = -k_D \cdot \dot{q}, \qquad (4)$$
$$F = \frac{1}{2} k_D \dot{q}^2, \quad Q_F = \frac{d}{dt}(\frac{\partial T}{\partial \dot{q}}) = 2 \cdot m \cdot \Omega_E \cdot cos\varphi \cdot \omega \cdot z_m \cdot cos\, \omega t.$$

Imagine that point (object) is placed in the force field above the Earth's surface and this point is forced to move vertically along Y axis. In accordance with the above information, the periodically compass force F_{com} will act on this point in the East-West direction. Let us consider the dynamics of the sensitive element motion.

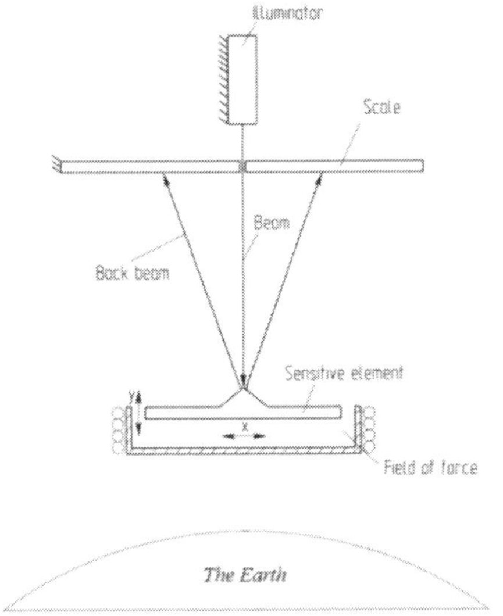

Fig. 2. Dynamics of the sensitive element motion.

We will obtain the expression (5) if the values of all variables are substituted in the Lagrange equations:

$$m\ddot{q} = -cq - k_D\dot{q} + H \cdot cos\, \omega t \qquad (5)$$

where $H = 2 \cdot m \cdot \Omega_E \cdot cos\varphi \cdot \omega \cdot z_m$

If we divide both sides of (5) by the mass, it will have a conventional view:

$$\ddot{q} + 2n\dot{q} + k^2\, q = h \cdot cos\omega t \qquad (6)$$

where $\frac{c}{m} = k^2, \quad \frac{k_D}{m} = 2n, \quad \frac{H}{m} = h$

The solution of this equation is the sum of homogeneous q_1 and particular solutions q_2.

The solution of the homogeneous equation q_1 is given by:

$$q_1 = A_1\, e^{-nt} sin\, k_1 t \qquad (7)$$

where $k_1 = \sqrt{k^2 - n^2}$.

Equation (7) determines the self-oscillation of the point (object).

In the amplitude form the equation (8) has the form:

$$q_2 = A_2 \sin(\omega t - \varepsilon) \qquad (8)$$

where A_2 – amplitude of forced oscillations; ε – phase angle of forced oscillations relative phase of the exciting force

Let us consider the forced oscillations (8), defined by the equation (8). It is necessary to replace in the expression (6) q to q_2, and to determine the first and second derivative. And if we do a number of required mathematical operations, a particular solution of the forced oscillations equation can be written as:

$$q_2 = \frac{h}{\sqrt{(k^2 - \omega^2) + 4n^2\omega^2}} \sin(\omega t - \varepsilon) \qquad (9)$$

To solve the proposed problem we consider the forced oscillations (9). The amplitude of these oscillations always direct to the East-West direction and is determined by the forced oscillations in the direction of the place vertical with the value of the amplitude H and the frequency ω. The frequency ω is necessary to select by varying the damping factor k_D, thus amplitude H. will be adjusted In accordance with the above information, the periodically complementary acceleration exists. Consequently the periodically motion with permanent direction East-West appears.

4 Conclusion

In this paper, theoretical substantiation and schematic solution of the practical implementation of the gravity-inertial gyrocompass without drift are presented. This gyrocompass can permanently monitor the East-West direction of the object regardless of time and it basing site.

References

1. G. Shvecov, et al *Gravitacionno-inercialnoe orientirovanie* (publishing TPU, Tomsk, Russia, 2004) (in Russian)
2. A.V. Taracenko, V.S. Dmitriev, T.G. Kostuchenko, *KORUS*, 83 (2004)
3. V. Dmitriev, Y. Britova, MATEC Web of Conferences **48**, 01005 (2016)
4. Yu. Britova, V. Dmitriev, T. Kostyuchenko, IOP Conference Series: Materials Science and Engineering **132(1)**, 012011 (2016)
5. V.S. Volkova, V.I. Kopytov, *5th Korea-Russia International Symposium on Science and Technology*, 29 (2001)
6. V.S. Ivanova, *8th International Scientific and Practical Conference of Students, Postgraduates and Young Scientists; "Modem Techniques and Technologies"*, 42 (2002)
7. P. Baranov, T. Nesterenko, E. Tsimbalist, E. Barbin, A. Koleda, S. Vtorushin, *IMEKO: Proc. 14th IMEKO TC10 Workshop on Tech.l Diag.* (2016)
8. V. Pavlov, *Teoriya giroskopa i giroskopicheskih priborov* (Sudostroenie, Leningrad, Russia, 1964) (in Russian)
9. G. Shvecov, USSR Patent № 4672218 (1991) (in Russian)

The tape winding current impact on the motor's torque curve

Antonina Dolgih[1,*], *Vladimir* Martemyanov[1], and *Ivan* Samodurov[1]

[1]Tomsk Polytechnic University, 634050 Tomsk, Russia

Abstract. The paper presents numerical simulations of the tape winding current impact on the motor's torque curve. The tape winding is a novel type of a motor's stator organization, allowing the large current flowing through the winding. Consequently the main flux of the permanent magnets changes. The estimation method of the developed torque dependence over the rotor and stator angular position is shown. Taking into account the control current effect the dependence is deformed, in particular it rotates relatively to the center.

1 Introduction

The torque curve is one of the main characteristics of the permanent-magnet torque motor. This curve shows the developed torque dependence from the current in the control winding. Generally, the torque curve is linear, without the demagnetizing effect of the winding current. It is possible with the rare earth magnets system. This statement relates to the torque motors, which are widely used in the modern devices [1-4]. However, the value of the control current in the tape winding torque motor [5] affects on the torque curve. It can be explained by the motor design, where the large current flowing is possible through the tape winding. This current generates the magnetic flux, which disfiguring the main flux of the motor.

Other important characteristics of the torque motor is the developed torque dependence over the rotor and stator angular position. This dependence allows to evaluate torque's pulsations during the rotor movement. The dependence can be determined considering the set of the torque curves over the different stator and rotor positions. In this set, the main motor's magnetic flux has the non-uniform deformation.

The purpose of this paper is to define the character of the torque curve distortion and to evaluate the changes of the developed torque from the control current.

In [6] it is shown that the torque of the tape winding motor is given by the expression

$$T = B_{\delta} \cdot \frac{r_{av} \cdot \Delta}{I_0} \cdot \frac{U}{R} \cdot D(x, y, I_0). \quad (1)$$

* Corresponding author: ivanovatonya@tpu.ru

© The Authors, published by EDP Sciences. This is an open access article distributed under the terms of the Creative Commons Attribution License 4.0 (http://creativecommons.org/licenses/by/4.0/).

Previously [6] assumed that flux density B_δ is the constant value in all subdomains, where the double integration is carried out. This is appropriate for the limited winding current $I = \frac{U}{R}$. However, at large currents the flux density depends on current $B_\delta(I)$.

Referring to (1), let us believe that the character of current lines distribution in the winding layers is independent from the flux density and current value: $D(x, y, I_0) = const.$ Besides, the geometric parameters are also considered as stable: $\delta = const$, $r_{av} = const$, and $\Delta = const$. Under these conditions (1) can be written as

$$T = k \cdot I \cdot B_\delta(I), \tag{2}$$

where $k = \frac{r_{av} \cdot \Delta}{I_0} \cdot D(x, y, I_0) = const.$

The equation (2) shows that the motor's torque is proportional to the winding current and depend on flux density in the air gap. And the flux density is the function of the current. Furthermore (2) can be written as

$$T = k_2 \cdot I \cdot \sum_{i=1, j=1}^{m,n} \left[D(x, y, I_0) \cdot B_\delta(I)\right]_{ij}, \tag{3}$$

where $k_2 = \frac{r_{av} \cdot \Delta}{I_0} = const.$

If the pole flux density is non-uniform, in particular distorted by the tape winding current, it is necessary to divide the tape surface under the pole on $(m \cdot n)$ certain subdomains. Each ij of these subdomains has the constant flux density, equal to the average value $(B_\delta)_{ij}$ for this subdomain. The double integral value $\left[D(x, y, I_0)\right]_{ij}$ should also be defined for that subdomain. This is shown in equation (3).

Further, the torque curve corresponding to the specified pole position over the tape winding pack can be defined as the dependence at the discrete current values

$$\left[F(I)\right]_x = \frac{T}{I} = k_2 \cdot \sum_{i=1, j=1}^{m,n} \left[D(x, y, I_0) \cdot B_\delta(I)\right]_{ij}. \tag{4}$$

2 Numerical simulation

The problem solution is carried out in COMSOL Multiphysics. COMSOL is an interactive environment for modeling and simulating of the scientific and engineering problems [7-9]. As a result of the numerical simulation the character of magnetic field distribution over the current flowing through the pack of plates in the diagonal direction is obtained. In this case the magnetic circuit is absent.

The blue arrows in Figure 1a show the magnetic flux density and the red lines represent the character of current flowing. Figure 1b shows the graphs of the magnetic flux density distribution over the width of the pack of 50 plates. The total current I_{tot} is equal to 2500 A. It defines as $I_{tot} = N \cdot I$, where N is the number of layers in the pack, I is the current flowing through one layer. Therefore, the generated in this case magnetic field will cause the significant distortion of the main motor's magnetic flux.

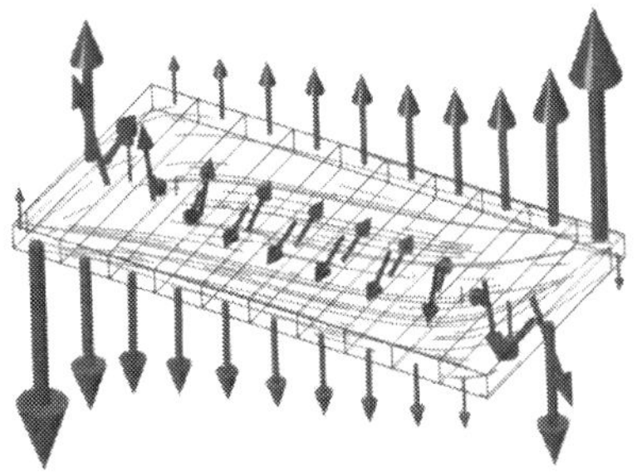

a)

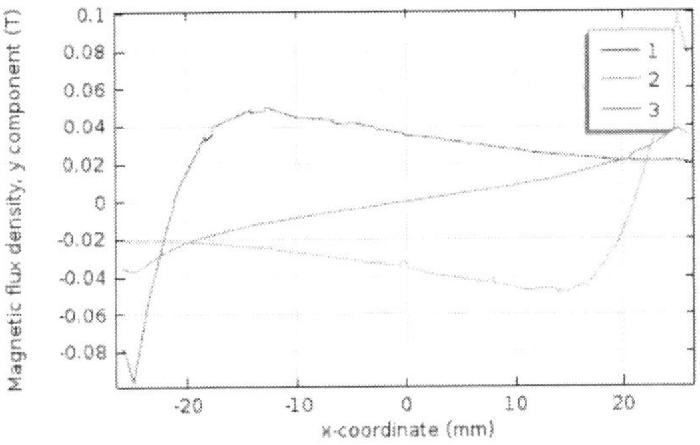

b)

Fig. 1. The character of magnetic field distribution: a - the pack of plates; b - the graphs of the flux density over the width: 1, 2 - at the ends of the pack; 3 - in the center of the pack.

The torque motor parametric model is built in T-flex CAD system. Figure 2a shows the tape winding fragment with the areas, where the flux density is defined. This is allowed to simulate the relative movement of the motor's magnetic system and the tape winding.

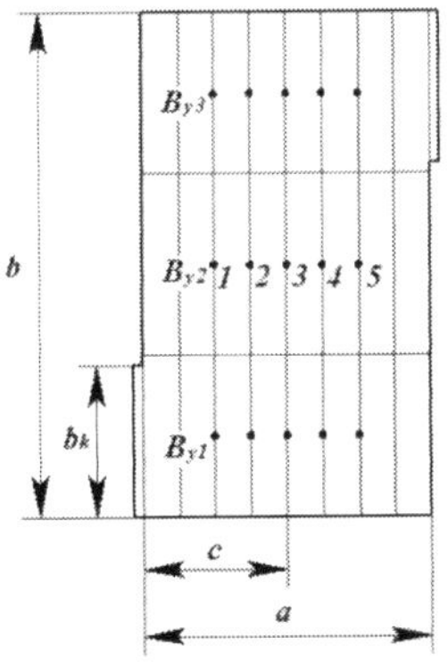

a)

b)

Fig. 2. The tape winding fragment: a - the subdomains of the tape winding pack; b - the character of the resulting magnetic flux density distribution at the position 1.

During the numerical simulation the flux density is determined in the positions indicated in Figure 2a by three variants:

I variant - Current flows through the tape winding. The magnetic circuit is absent.

II variant - Current flows through the tape winding. The motor's rotor is made of ferromagnetic material, without permanent magnets.

III variant - Current flows through the tape winding. The rotor is represented by neodymium magnets and ferromagnetic poles.

If the current through the tape winding is absent, the motor's magnetic field will have the uniform magnetic field with the flux density about 0.4 T. If the current flows through the tape winding (variant III), the main magnetic flux will deform. Figure 2b shows the character of the resulting magnetic flux density distribution at the position 1.

The obtained results allow to define the flux density distribution in the air gap. These data are available at the different pole positions. In this case, the five positions are selected.

3 Results and Discussion

Based on (3), results of numerical simulation and the previously obtained values of double integral for all selected subdomains of the tape winding pack the developed motor's torques over the five selected rotor positions are calculated. The results are shown in Figure 3 by the dependences for two values of the total current, flowing through the tape.

The curves 1 and 3 represent the dependences without the effect of control current, flowing through the tape winding. The curves 2 and 4 take into account the effect of that current.

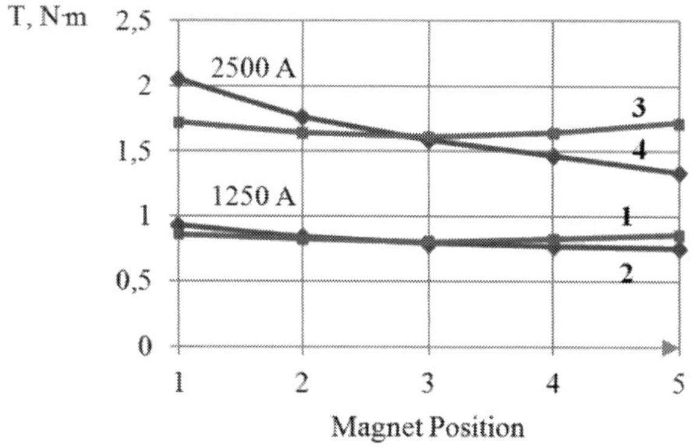

Fig. 3. The motor's torque over the angular movement.

4 Conclusion

From the graphs in Figure 6 it follows that taking into account the effect of the control current the dependence of the developed torque over the permanent magnet position is deformed. In particular, there is the rotation of the dependence relatively to it center. The reversal current cause the rotation in the other direction.

At the total current of the tape winding 2500 A the deformation reaches (25-30) %; at 1250 A – 12 %. In some technical devices, such parameters are permitted. If it is necessary the dependence can be corrected for example by the additional side cuts inside the tape winding plate [6].

References

1. J. F. Gieras, *Permanent Magnet Motor Technology: Design and Applications* (Marcel Dekker, Inc., New York, 2010)
2. Dr. Fritz Faulhaber GmbH & Co. KG, *Technical information* (2016)
3. R. Crowder, *Electric Drives and Electromechanical Systems: Applications and Control* (Elsevier Science & Technology Books, London, 2006)
4. C. C. Tsai, S. C. Lin, H. C. Huang, Y. M. Cheng, Mechatronics **19**, 29 (2009)
5. V. Martemjanov, A. Dolgih (Ivanova), *SIBCON*, 6693632 (2013)
6. A. Dolgih, V. Martemyanov, MATEC Web of Conf. **48**, 01004 (2016)
7. V. Ogay, P. Baranov, A. Stepankova IOP Conference Series: Materials Science and Engineering **66**, 012009 (2014)
8. P. Baranov, V. Baranova, MATEC Web of Conf. **48**, 03001 (2016)
9. E.V. Golodnykh, V.N. Borikov, *SIBCON*, 6693635 (2013)

Frequency-domain imaging algorithm for ultrasonic testing by application of matrix phased arrays

Dmitry Dolmatov[1,*], *Yana* Salchak[1], and *Roman* Pinchuk[2]

[1]Tomsk Polytechnic University, 634050 Tomsk, Russia
[2]ACS-Solutions GmbH, Science Park 2, 66123 Saarbrücken, Germany

Abstract. Constantly increasing demand for high-performance materials and systems in aerospace industry requires advanced methods of nondestructive testing. One of the most promising methods is ultrasonic imaging by using matrix phased arrays. This technique allows to create three-dimensional ultrasonic imaging with high lateral resolution. Further progress in matrix phased array ultrasonic testing is determined by the development of fast imaging algorithms. In this article imaging algorithm based on frequency domain calculations is proposed. This approach is computationally efficient in comparison with time domain algorithms. Performance of the proposed algorithm was tested via computer simulations for planar specimen with flat bottom holes.

1 Introduction

Methods of nondestructive testing are used extensively in aerospace industry due to high requirements for materials and systems quality. Ultrasonic testing has become one of the most promising techniques for this purpose considering its accuracy and low costing [1].

Accuracy of the measurement can be influenced by the type of the chosen transducer. Ultrasonic phased array probes are most common for industrial applications. Due to the quantity of the elements it provides diversity of scanning laws that is important for the precise imaging. In general, phased array transducers can be divided into two groups: linear and matrix phased arrays. The main advantage of matrix phased array transducers is that it can be used for 3-D reconstruction. Application of linear arrays is limited for this purpose [2]. Further progress in ultrasonic imaging is determined by development of fast imaging algorithms within the application of matrix phased arrays.

There are two basic types of imaging algorithms: frequency-domain and time-domain. However, it is known that frequency-domain algorithms have better computational speed in comparison with time-domain type [3]. By that reason it is challenging to develop a new three-dimensional imaging algorithm based on calculations in frequency domain. In this article the algorithm based on so-called Stolt transformation is proposed [4]. The solution is formed in accordance with the results of the previous research for two-dimensional

* Corresponding author: dolmatovdo@tpu.ru

© The Authors, published by EDP Sciences. This is an open access article distributed under the terms of the Creative Commons Attribution License 4.0 (http://creativecommons.org/licenses/by/4.0/).

geometry and single element case [5]. Thereby, frequency domain algorithm based on Stolt mapping should be extended for application of full matrix capture technique in order to build three-dimensional reconstruction.

2 Materials and methods

The proposed algorithm consists of three main steps. First is application of full matrix capture technique. Thus, the function is to be changed for separate transmitter and receiver elements. The second step is extrapolation of the wave field to the depth of the boundary between water and the specimen. And the final step is three-dimensional Stolt mapping of scattered data.

Full matrix capture is the technology that involves collecting of scattered data from the full set of receiver/transmitter pairs. In this case the measured wave field can be described by the function $F(t, x_{tr}, y_{tr}, z_{tr}, x_r, y_r, z_r)$. This function depends on time and coordinates of the transmitter and the receiver. In this article application of a plane matrix phased array is proposed. Due to this fact one of the coordinates of transmitter and receiver will be the same. In this case function $F(t, x_{tr}, y_{tr}, x_r, y_r, z)$ can be reconsidered. It is to be transformed to $F(t,x,y,z)$, where x and y variables are found according to the following correlations:

$$x = \frac{x_{tr} + x_r}{2} \tag{1}$$

$$y = \frac{y_{tr} + y_r}{2} \tag{2}$$

The next step is wave field extrapolation to the boundary depth between water and the specimen. Firstly, it is necessary to calculate three-dimensional fast Fourier transform of the scattered data:

$$P(\omega, k_x, k_y, z) = 0.125\pi^{-3} \int\int\int_{-\infty}^{\infty} F(t, x, y, z) e^{-i(k_x x + k_y y - \omega t)} dx\, dy\, dt\,, \tag{3}$$

where k_x,k_y are wavenumbers, ω is a temporal frequency and $0.125\pi^{-3}$ is normalization constant.

After that the wave field extrapolation to the desired depth is calculated:

$$P(\omega, k_x, k_y, z + \Delta z) = P(\omega, k_x, k_y, z) e^{ik_z \Delta z}\,. \tag{4}$$

The last step is Stolt mapping of extrapolated wave field. The final image of the specimen is derived according to the equation:

$$I(x, y, z) = \int\int\int_{-\infty}^{\infty} \frac{c_l}{2} \frac{k_z}{\sqrt{k_x^2 + k_y^2 + k_z^2}} P(k_x, k_y, k_z) e^{-i(k_x x + k_y y + k_z z)} dk_x\, dk_y\, dk_z\,, \tag{5}$$

where c_l is a velocity of longitudinal waves in specimen. It needs to be noted that before the implementation of this equitation it is vital to change the variable. It means that temporal frequency ω is to be changed according to the wavenumber k_z :

$$\omega(k_x, k_y, k_z) = \frac{c_l}{2} \sqrt{k_x^2 + k_y^2 + k_z^2} \tag{6}$$

3 Results and discussion

Computer simulation is used for verification of the proposed algorithm. This simulation is built using special software CIVA 2015. This is a powerful and versatile tool for computer simulations of ultrasonic inspections [6].

The simulated specimen is a planar steel block within several flat bottom holes. The diameter of each hole is 1 mm. Simulation setup is shown in Figure 1. Each flaw is marked by a corresponding letter. The position of each hole can be determined by x and y coordinates. Further, each defect has its own particular depth. Hereby, the position is to be described by three parameters (x,y,z), where x and y are coordinates of the flat bottom hole and z is its depth (all values are in mm). In total there were five flat bottom holes created for simulation. The corresponding coordinates are: A(-5;5;35), B(-2.5;2.5;30),C(0;0;15), D(2.5;-2.5;20), E(5;-5;25).

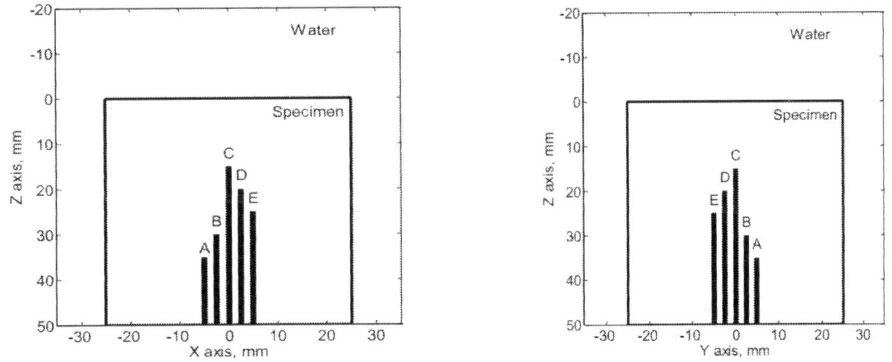

Fig. 1. Simulation setup.

In order to achieve reliable results, it is important to choose appropriate parametres of the transducer. A sixty four element phased matrix array was chosen as a transducer with the frequency of 5MHz. Width and length of each element of the matrix is 0.5 mm. Spacing between elements is 0.1 mm. As for the inspection settings, the step of 2 mm was chosen for accurate simulation.

Proposed algorithm is post processing technique for imaging of received acoustic signals. Thus, the results of the simulation were used as input data. For the estimation of the proposed algorithm efficiency Matlab 2016a (The MathWorks, Natick, MA) was implemented.

A common way of three-dimensional ultrasonic imaging is visualization of the back-scattered amplitude maximum corresponding to each measuring point along the depth. This approach was used to demonstrate the results of algorithm implementation given in Figure 2.

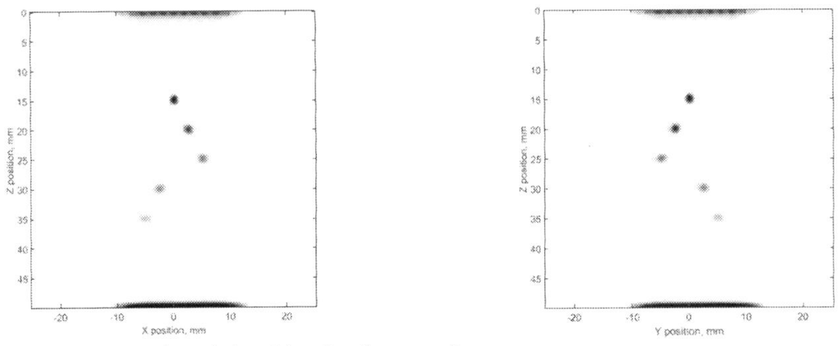

Fig. 2. The results of algorithm implementation.

Experimental results demonstrate that the proposed frequency-domain algorithm was improved and can be applied for three-dimensional imaging within the matrix phased array application. Accurate and precise measurements results were received for each of the flaws by testing the suggested algorithm.

Another important parameter that can influence the reliability of imaging is positioning precision of the system. Robotic instruments are generally applied in order to achieve higher precision. However, the calibration of these systems is quite challenging. In our previous research we suggest a novel approach of noncontact calibration that can be implemented for this purpose [7, 8].

4 Conclusions

Computer simulation in CIVA was made to verify the proposed algorithm. This is a novel frequency-domain 3-D imaging algorithm for ultrasonic testing with matrix phased arrays. The achieved results are validated and suggest solution that can be applied for effective immersion ultrasonic testing of planar specimens within the application of matrix phased arrays.

The reported study was partially supported by the Governmental program "Science", research project No. 11.6342.2017/BCh.

References

1. G. Scott, C. M. Scala, NDT international **15.2**, 75 (1982)
2. O. Oralkan, A. S. Ergun, C.H. Cheng, J.A. Johnson, M. Karaman, T,H. Lee, B.T. Khuri-Yakub, IEEE T Ultrason Ferr **50(11)**, 1581 (2003)
3. A.J. Hunter, B.W Drinkwater, P.D Wilcox, IEEE T Ultrason Ferr **55(11)**, 2450 (2008)
4. R.H Stolt, Geophysics **43.1**, 23 (1978)
5. D. Dolmatov, V. Abramets, Matec Web Conf **48,** 3004 (2016)
6. P. Calmon, S. Mahaut, S. Chatillon, R. Raillon, Ultrasonics **44**, 975 (2006)
7. D. Dolmatov, V. Zhvyrblya, G. Filippov, Y. Salchak, E. Sedanova, IOP Conf Ser Mater Sci Eng **135(1),** 12010 (2016)
8. G. Filippov, V. Zhvyrblya, S. Sharavina, Y. Salchak, Matec Web Conf **48,** 3002 (2016)

Design of optoelectronic system for optical diffusion tomography

Igor Erakhtin[1,*], *Aleksandr* Aristov[1], *Anna* Novoseltseva[2], and *Viktor* Sukhanov[3]

[1]Tomsk Polytechnic University, 634050 Tomsk, Russia
[2]Worcester Polytechnic Institute, MA 01609-2280 Worcester, USA
[3]Siberian Branch, Russian Academy of Sciences, 630090, Novosibirsk, Russia

Abstract. This article explores issues connected with the circuit design of a device for optical diffusion tomography, which we are currently designing. We plan to use the device in experimental studies for the development of a faster method of brain hematoma detection. We reviewed currently existing methods for emergency diagnosis of hematomas, primarily the Infrascanner model 2000, for which we identified weaknesses, and outlined suggestions for improvements. This article describes the method of scanning tissues based on a triangulated arrangement of sources and receivers of optical radiation, and it discusses the optoelectronic system that implements that principle.

1 Introduction

Brain injuries are often accompanied by intracranial hematomas, the presence of which is life-threatening. Therefore, quick detection of hematoma, including its size and location, is important for effective medical care. Today, there are two primary methods for the detection of intracranial hematomas: magnetic resonance and computed tomography. Carrying out an examination of patients by these methods is not always possible, since they are not always accessible. Therefore, creating a portable device that is able to identify the presence of intracranial injuries is important. A portable device can be achieved using spectroscopy in the optical range of wavelengths.

2 Problem definition

Nowadays, there is a device on the market, Infrascanner model 2000 [1] developed by InfraScan company. It is a portable detector of intracranial hematomas, operating in the near infrared range (808 nm). According to the experimental investigations [2, 3], the Infrascanner model 2000 has a high diagnostic sensitivity. The company published that the probability of detecting the presence of hematomas is 91% (assuming that the occurrence of hematoma is not more than 2.5 cm from the brain cortex and is not larger than 3.5 cm^3). Also, the device does not determine the parameters of the hematoma (size, depth of bedding, exact localization) which are very important for resuscitation and rehabilitation.

* Corresponding author: igor_e_k_94@mail.ru

© The Authors, published by EDP Sciences. This is an open access article distributed under the terms of the Creative Commons Attribution License 4.0 (http://creativecommons.org/licenses/by/4.0/).

The operation principle of the Infrascanner model 2000 is based on a differential method, i.e. the identification of hematomas is done by comparing the optical properties of symmetrical sections on the head at the given points [1]. Thus, there is a high probability of missing small or deeply bedded hematomas outside of these zones. In order to eliminate these disadvantages, improvements of the method and the device which will help in collecting qualitative and quantitative data (e.g. hematoma parameters) are necessary.

3 Results

The Infrascanner devices use only one radiation source and a receiver, which limits its diagnostic capabilities. An increase of the number of sources and photodetectors may help to improve hematoma parameters detection. We assume that the optimal scheme is a scheme with radiation sources and detectors located at the vertices of an equilateral triangle (Figure 1).

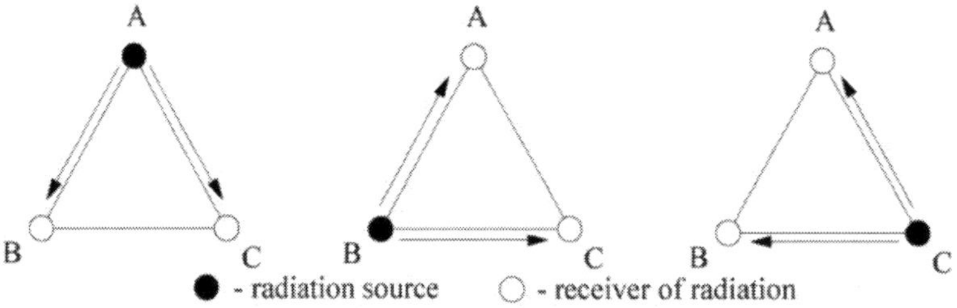

Fig. 1. Scheme of triangulation method.

As shown in Figure 1, the radiation source is located at point A and the receivers at points C and B, respectively, this configuration is used for the first measurement. Then the source moves to point C, and the receivers are located at points A and B during the second measurement. Finally, the source moves to point B, and the receivers are at points A and C during the third measurement. As a result of these measurements, we can detect shape and location of hematoma.

Additionally, two-wave spectroscopy can be used to increase the examination sensitivity. A hematoma is a clot of coagulated blood. The earlier works [4] have shown that there is a significant spectral difference for clotted blood and incoagulated blood in the red region of the spectrum. Therefore, the measurement can be carried out in the red (about 640nm) and infrared (about 805 nm) regions of the spectrum. A number of issues relating to the development of such systems for hematoma detection were considered in publications [5, 6].

In view of the device requirements and research objectives, it was proposed to construct an experimental model, as described below.

The system sequentially reads the information from two photodetectors that collect a red and an infrared radiation transmitted through the sample. Therefore, the device consistently generates two pulses of red and infrared light with a minimum time delay. Then, these two radiations that passed through the examined biological tissue are measured and the results are saved. Further, a cycle of measurements is carried out for the subsequent combinations of radiation sources and receivers. Then the information is analyzed and the results are displayed on an indicatory device. On the basis of the abovementioned, one of the best solutions is a creation of the device based on a microcontroller, that generates pulse sequences to measure parameters of the signal, data analysis and its display on the indicatory device.

We use laser diodes as the radiation sources since they have high-speed performance

 DOI: 10.1051/matecconf/201710201016

and sufficient emissive power. We use a voltage-to-current converter as a supply source for an adjustment of the laser diode current. For the photo-sensor element, it is better to use photodiodes since in comparison with other similar photo-sensor elements the photodiode has high integral sensitivity, a wide range of wavelengths, the linear dependence of a photoelectric current of the diode from emissive power, as a result – simpler data analysis. As the signal from the photodiode is very small, it needs to be amplified to an appropriate level. In addition, an ambient illumination interferes the signal on the photodiodes. In order to eliminate this interference, the method of measuring the backlight voltage and subtracting it before supplying the laser diodes can be used. Thus, it will help to get rid of a constant component of backlight voltage, and digital filtering will be used for elimination of 50 Hz and 100 Hz signals. After filtering, the signal will be amplified to the level of analog-to-digital converter (ADC) operation. Since head tissues in different regions of a head have significant variance of the optical properties, the final amplifier needs an automatic gain control. For the realization of the triangulation method, there are should be three receiving and transmitting channels. Schematic diagram of our optoelectronic system (which includes abovementioned modifications) is shown in Figure 2.

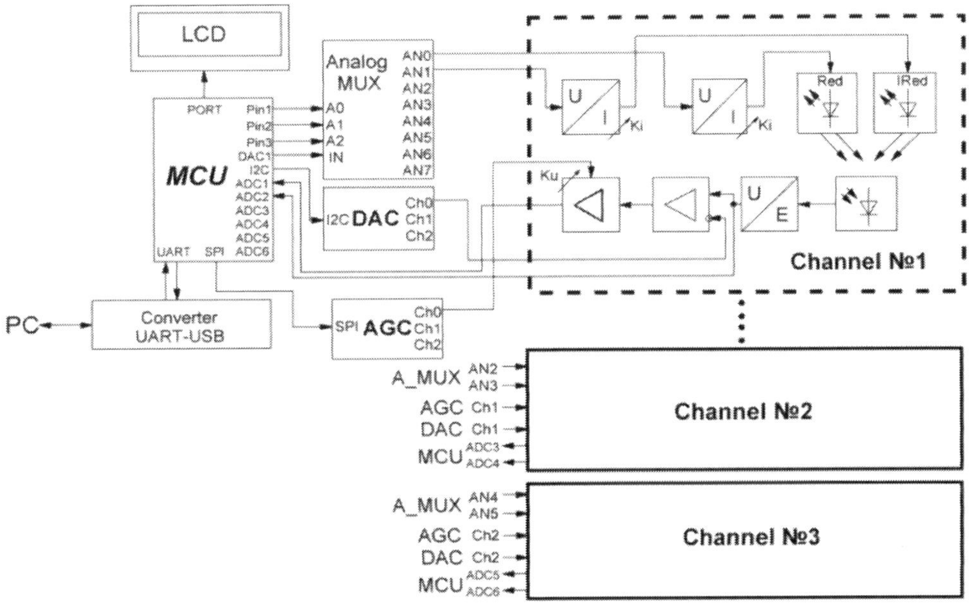

Fig. 2. Schematic diagram of an optoelectronic system for biological tissues investigation. *MCU* - microcontroller unit; *AGC* - automatic gain control; *DAC* - digital-to-analog converter; *LCD* - liquid-crystal display; *PC* - personal computer; *ADC* - analog-to- digital converter.

The role of MCU executes the microcontroller STM32F407VG: it organizes the work of all blocks and performs data analysis. Analog multiplexer is used for the intensity modulation of the probing radiation by commutation of the DAC voltage to the voltage-to-current converters, which are connected to the laser diodes. The transimpedance amplifiers are used for the signal conversion from the photodiodes. The differential amplifier eliminates an ambient light influence by applying the voltage from the DAC module to the inverting input of the differential amplifier. The applied voltage magnitude equals to the signal value measured using ADC of microcontroller at the moments when the lasers are turned off. The output stage consists of the inverting amplifier based on the operational amplifier with a digital resistor in the feedback of the amplifier. The digital resistor is used for the realization of the automatic gain control. The AGC unit is an assembly of the digital resistors located in the feedback of the operating amplifier.

4 Conclusions

We developed the schematic diagram of an optoelectronic device intended to eliminate the disadvantages we found in the Infrascanner model 2000. We designed the schematic circuit of an optoelectronic device and created the prototype of the single-channel device. The operation units and device control algorithms will be debugged using the prototype. Currently, the automatic gain control of the output stage has been debugged. The program code for the MCU for coordinated operation of the circuit blocks has been written. The digital filtering of the signal was implemented using a 1st order Butterworth low-pass filter. The connection of the prototype with the PC was completed using UART. The PC application was written for convenient testing of the device.

In the future, for implementation of the triangulation method, the number of channels of the prototype will be increased to three. The device will be tested on the model objects. On the basis of these tests, we plan to develop algorithms, which will allow the detection of hematoma and its parameters, namely: size, depth, and location.

References

1. *The Infrascanner Model 2000*, URL: http://infrascanner.com/wp-content/uploads/2014/12/IFS-Brochure.1.9.15.pdf
2. S. Bressan, M. Daverio, et al., Childs Nerv Syst **30,** 477 (2013)
3. C.S. Robertson, E.L. Zager, et al., Journal of Neurotrauma **27**, 1597 (2010)
4. A.P. Novoseltseva, K.A. Timchenko, A.A. Aristov, D. Kustov, I.A. Larioshina, Journal of Physics: Conference Series **671,** 012002 (2016)
5. K.A. Timchenko, A.A. Aristov, I.S. Musorov, T.G., Evtushenko, *Proceedings of the 15th International Conference of Young Specialists on Micro/Nanotechnologies and Electron Devices,* 319 (2014)
6. A.P. Novoseltseva, K.A. Timchenko, A.A. Aristov, IOP Conf. Ser: Mat.Sc. and Eng. **93**, 012022 (2015)

Sensor module for testing magnetometric borehole inclinometers under field conditions

Anatoly Gormakov[1,*], *Mikhail* Kharitonov[2], and *Andrey* Prygov[1]

[1]National Research Tomsk Polytechnic University, 634050 Tomsk, Russia
[2]Research and Development Centre 'Polyus', 634050 Tomsk, Russia

Abstract. The paper presents the testing procedure and the test unit for borehole inclinometers using a reference system of orientation incorporated in the sensor module. Computational algorithms are developed for azimuth, zenith and apsidal angles. The estimated error is obtained for the detection of zenith angles. The sensor module uses the Honeywell HMS1053, a digital resistance magnetometer and the Colibrys MS9000 accelerometer. The size and weight of the test unit are rather small that provides its easy transportation. The test unit is universal as it can be used for inclinometers with different casing diameters. The cost of the test unit is much lower than that of the stationary one. Easy servicing reduces the labor content during the inclinometer in situ testing.

1 Introduction

The outer space exploration begins with the first stages of launch site construction and ends with the borehole drilling aimed at the investigation of the structure and composition of the surface and deep layers of cosmic bodies [1].

The metrological support of the inclinometer survey instruments is provided by the metrological installations for calibration and testing borehole inclinometers of the types USI-2, UPI-1, UKI-2, UNP-3 and some others [2-6]. These installations can be used only in steady-state conditions of the inclinometer manufacturer or in laboratories of geophysical companies. An operational test of inclinometers has not been provided *in situ* until present time. In [7-8] the authors proposed the improved testing procedure and the test unit for borehole inclinometers.

2 Test unit and procedure for testing borehole inclinometers

The procedure for testing borehole orientation system is presented in Figure 1. The sensor module *5* which incorporates the reference orientation system is attached to the inclinometer casing *1* using the frame (prism) *2*, brackets *3*, and pressure screw *4*.

The sensor module allows mounting inclinometers of different casing diameters. The sensor module is mounted to the inclinometer casing *via* the prismatic surface of the frame

* Corresponding author: gormakov@tpu.ru

© The Authors, published by EDP Sciences. This is an open access article distributed under the terms of the Creative Commons Attribution License 4.0 (http://creativecommons.org/licenses/by/4.0/).

2. This construction provides the alignment of sensitivity axes of the sensor module *5* and the inclinometer casing *1*. An arbitrary spatial attitude is then given to the inclinometer casing. The signals from the inclinometer casing and the sensor module enter the data collection units *6* and *7*, respectively. The obtained results are interpreted by the computer *8* and displayed on the monitor, so the performance of the sensor module can be then evaluated. In case the sensor module matches the accuracy requirements, it is put into operation. Otherwise, it should be sent to a repair facility.

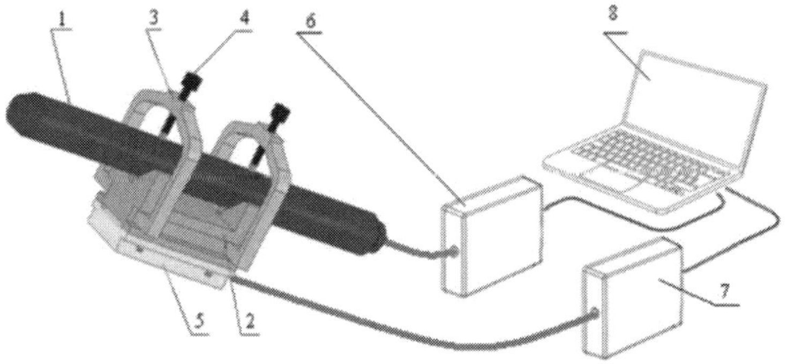

Fig. 1. Schematic drawing of inclinometer test unit using the sensor module.

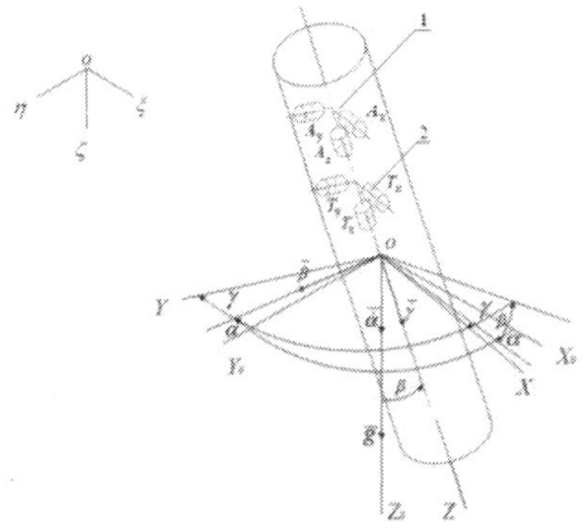

Fig. 2. Sensor module of the reference system of orientation.

In the related references [11–14], it was observed that in years ahead, the borehole orientation systems based on MEMS [15] accelerometers and resistance magnetometers will be widely used for bottom-hole inclinometers thanks to their operability under the vibration, shaking and high-temperature conditions.

The proposed sensor module uses the Honeywell HMS1053, a digital resistance magnetometer [16-17] and the Colibrys MS9000 accelerometer [18].

3 Sensor module configuration

The general configuration of the sensor module is presented in Figure 2. The accelerometer unit *1* (A_x, A_y, A_z) and the magnetometer unit *2* (T_x, T_y, T_z) are placed along each of the axes of *OXYZ* coordinate system connected to the sensor module housing. The linear accelerometers A_x, A_y, A_z measure the trimetric projection of apparent accelerations a_x, a_y, a_z which is connected to the sensor module housing.

According to Figure 2, the axis $O\zeta$ of the reference coordinate system $O\xi\eta\zeta$ is directed downward, while the axes $O\xi$ and $O\eta$ are horizontal, $O\xi$ being lied in a plane of the astronomical meridian and directed northward. The axes of $OX_0Y_0Z_0$ coordinate system always originating from the point O of the test unit are parallel to the respective axes of $O\xi\eta\zeta$ coordinate system.

In order to obtain the computational algorithms for azimuth angle, let us consider three sequential rotations of $OXYZ$ coordinate system relative to $OX_0Y_0Z_0$ coordinate system, respectively at azimuth angle α, zenith angle β, and apsidal angle γ.

The vector components for gravitational acceleration g on the axis of $OXYZ$ coordinate system can obtained from Eq. (1).

$$\left.\begin{aligned} a_x &= -g \cdot sin\beta \cdot cos\gamma; \\ a_y &= g \cdot sin\beta \cdot sin\gamma. \end{aligned}\right\} \tag{1}$$

Using the system of equations (1), the main value of zenith angle β^* is written as

$$\beta^* = arcsin\sqrt{\frac{a_x^2 + a_y^2}{g^2}}, \tag{2}$$

the apsidal angle γ can be obtained from

$$\gamma = -arctg\frac{a_y}{a_x}, \tag{3}$$

and azimuth angle α is calculated from

$$\alpha = arctg(\frac{T_x \cdot sin\gamma - T_y \cdot cos\gamma}{T_x \cdot cos\beta \cdot cos\gamma + T_y \cdot cos\beta \cdot sin\gamma - T_z \cdot sin\beta}), \tag{4}$$

where T_x, T_y, T_z are the signals from the respective magnetoresistive sensors.

4 Error estimation for zenith angle detection

The error estimation of the sensor module when measuring the zenith angle is provided by the theorem for functions of several variables. Thus, the error is found from where Δa_x, Δa_y are the measurement errors of accelerometers placed on X- and Y-axes; Δg is the error of gravitational acceleration.

$$\Delta\beta = (\left|\frac{a_x}{g^2 \cdot \sqrt{\frac{a_x^2 + a_y^2}{g^2}} \cdot \sqrt{1 - \frac{a_x^2 + a_y^2}{g^2}}}\right| \cdot |\Delta a_x|) + (\left|\frac{a_x}{g^2 \cdot \sqrt{\frac{a_x^2 + a_y^2}{g^2}} \cdot \sqrt{1 - \frac{a_x^2 + a_y^2}{g^2}}}\right| \cdot |\Delta a_y|) + \tag{5}$$
$$+(\left|\frac{a_x}{g^3 \cdot \sqrt{\frac{a_x^2 + a_y^2}{g^2}} \cdot \sqrt{1 - \frac{a_x^2 + a_y^2}{g^2}}}\right| \cdot |\Delta g|)$$

The error calculations showed that the zenith angle error did not ecxeed 0,05% at zenith angles ranging from 0° to 85° and all possible values of apsidal angle γ. At zenith angle β bing close to 90°, the error is 0,17%. It should be noted that the obtained errors are maximum errors which, however, can hardly be expected to appear.

5 Conclusions

The research investigations resulted in the construction of the flow sheet of the reference borehole orientation system; selection of sensor elements; development of computational algorithms for orientation angles; and estimation of measurement errors for zenith angle. The proposed test unit can be used for testing borehole inclinometers in the field conditions. The use of the Honeywell HMS1053, a digital resistance magnetometer and the Colibrys MS9000 accelerometers will allow the researchers to implement a safe structure of the sensor module possessing the small size and weight.

References

1. *Burenie v kosmose*, URL : http://byrim.com/burenie/13.html
2. E.V. Golodnykh, V.N. Borikov, *SIBCON*, 6693635 (2013)
3. Z.G. Gareishin, *Concepts of assembling metrological setups of borehole electronics orientation* (Ural-Geo Centre, Ufa, 2006)
4. A.N. Golikov, A.N. Gormakov, Kontrol'. Diagnostika. **Spec. issue**, 42 (2011) (in Russian)
5. *Ustanovka UNP-3 KSVSh*, URL: www.oskbp.ru/index.php?id=19 (in Russian)
6. *Gyroscopic inclinometr «IGM33/42*, URL: www.trend.miass.ru/prod02_e.htm
7. A.N. Gormakov and et al, RF Patent (2010) (in Russian)
8. A.N. Gormakov, I.A. Ul'yanov, V.G. Tkachev, Vest. Nauki Sibiri. **2,** 101 (2014) (in Russian)
9. J. Bojja and et al, IEEE Sens. J. **16**, 8, (2016)
10. V.S. Volkova, V.I. Kopytov, *KORUS 2001,* 29 (2001)
11. V.S. Ivanova, *MTT 2002,* 59 (2002)
12. *Solod-state MEMS inclinometer*, URL: www.radiant.su/files/images/sq/sq-si-360da%20(sq-six).pdf
13. *Digital inclinometer system*, URL: www.rstinstruments.com/Digital% 20MEMS% 20Inclinometer.html
14. *Arduino inclinometer,* URL: http://appsmagz.us/arduino/arduino-inclinometer
15. P. Baranov, and et al, *IMEKO* (2016)
16. A. Borisov, Comp. & Technol., **7**, 56 (2006)
17. *1,2 and 3 axis magnatic sensors,* URL: www.electronshik.ru/pdf/pdf/h/ hmc105x.pdf
18. Yu. Ponomarev, Comp. & Technol., **10**, 34 (2015)

Total harmonic distortion of an asymmetric quasi-sinusoidal current

Vitaliy Grebennikov[1,*], *Irina* Ermoeva[1], and *Evgeniy* Yaroslavtsev[1]

[1]Tomsk Polytechnic University, 634050 Tomsk, Russia

Abstract. This paper presents the correlation of factors that determine the quality of asymmetric quasi-sinusoidal output current and dynamic losses in the switches of the current generator circuit. The operating mode of the generator was obtained, especially of its power circuit elements, where combination of acceptable output current quality with relatively small dynamic losses in switches and mass-dimensional parameters of the inductor are provided. Achieved results can be used in designing this type of generators.

1 Introduction

Quasi-sinusoidal current is widely used in electrochemical and electrophysical technologies [1]. One of the generator circuits, which provides this type of current, is described in [2] and shown in Figure 1. The shape of half-waves of the generator output current which is shown in Figure 2 is quasi-sinusoidal because of its formation principle. This shape can be approximated to an ideal one by reducing the ripples of current and increasing the amount of switching cycles. It causes the increasing of power transistors switching frequencies and, as a result, the increasing of dynamic losses in switches of the current generator. Therefore, there is a problem how to choose an effective operating mode of generator power circuit elements where an acceptable output current quality combines with relatively small dynamic losses in switches and with small mass-dimensional parameters of the inductor.

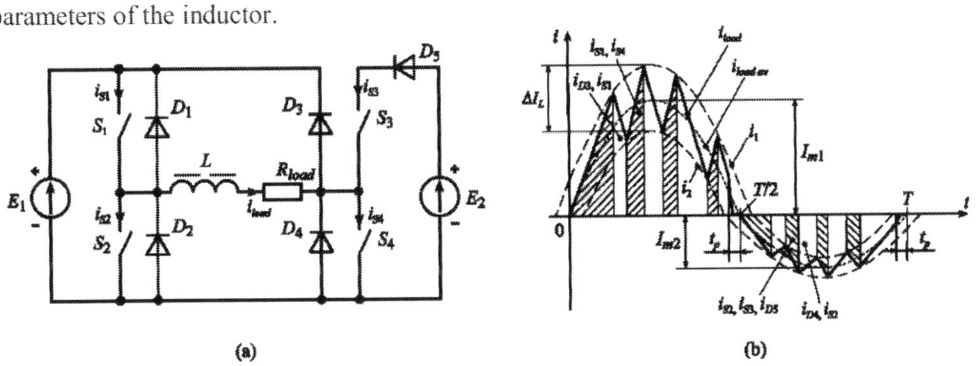

Fig. 1. Current generator circuit (a) and the current waveforms are explained of the circuit operation (b).

* Corresponding author: grebennikovvv@tpu.ru

© The Authors, published by EDP Sciences. This is an open access article distributed under the terms of the Creative Commons Attribution License 4.0 (http://creativecommons.org/licenses/by/4.0/).

2 Materials and methods

Total harmonic distortion (THD) is used to evaluate the output current quality of the generator. The value of this factor for asymmetric quasi-sinusoidal current depends on two parameters:

- non-ideality of each one of the half-waves (instead of an ideal sinusoidal shape) which are formed by half-waves current converter (as a result of formation principle);
- asymmetry of half-waves $S = I_{m1}/I_{m2}$.

Both of these parameters cause the appearance of higher harmonics, and it becomes the reason for harmonic factor to get definite values, which are different from zero. Let us examine a separate influence of the mentioned parameters on the THD of the asymmetric quasi-sinusoidal current, conducting the necessary research using OrCAD simulation. The definition of total harmonic distortion is set for first 40 harmonics in the THD calculation directive.

The circuit shown in Figure 1 was used for simulation. Parameter S of the asymmetric quasi-sinusoidal current was equal 1 ($S = 1$) to exclude the influence of the asymmetry on the THD. The output current has a symmetric quasi-sinusoidal shape. To make modeling results objective, the THD of this quasi-sinusoidal current is equal to 12%.

The following notations are used: ΔI_L – required value of current ripples; $K_{rip} = \Delta I_L/I_m$ – ripple factor of load and inductor current; $U^* = I_m \cdot R_{load}/E = U_{m\,load}/E$ – normalized value of the average amplitude of the load voltage with respect to the supply voltage (then – the normalized output voltage amplitude); $U_{m\,load}$ – the average amplitude of the voltage across the load; $\tau = L/R_{load}$ – the time constant; $\delta = T/\tau$ – the damping ratio of the transient showing how many times the period of a sine wave is higher than the time constant; T – period of the output current; f – frequency of the output current; N – amount of switching cycles of switches in half-period of formed quasi-sinusoid.

3 Results

The results of the simulation are shown in Figure 2 and Figure 3. They are presented as characteristics of THD and N, and they depends on δ at various values of K_{rip} and U^*.

Let us view these characteristics more in details. When δ is small THD has the biggest values on every graphs, and when $K_{rip} = 0{,}3$ and $K_{rip} = 0{,}4$ they are above permissible maximum limits 12%. It is explained by respectively big ripples of the formed current and comparatively small number of switching cycles of switches N that appear after long

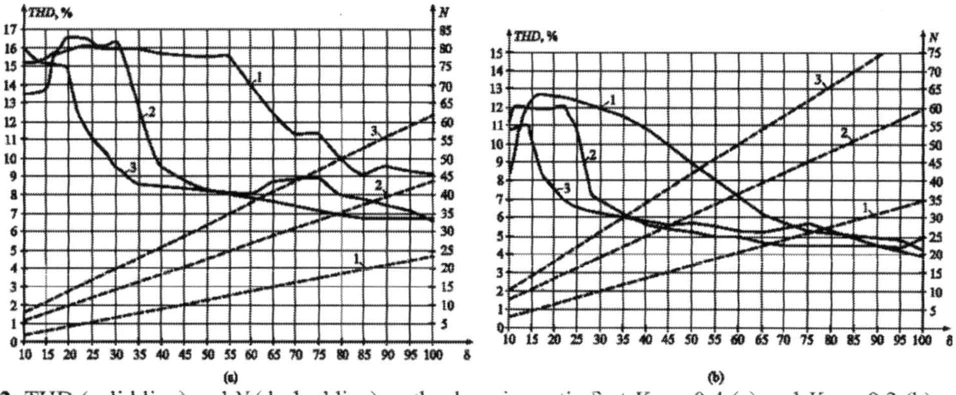

Fig. 2. THD (solid line) and N (dashed line) vs the damping ratio δ at $K_{rip} = 0{,}4$ (a) and $K_{rip} = 0{,}3$ (b) for different U^*: 1 – $U^* = 0{,}8$; 2 – $U^* = 0{,}5$; 3 – $U^* = 0{,}2$.

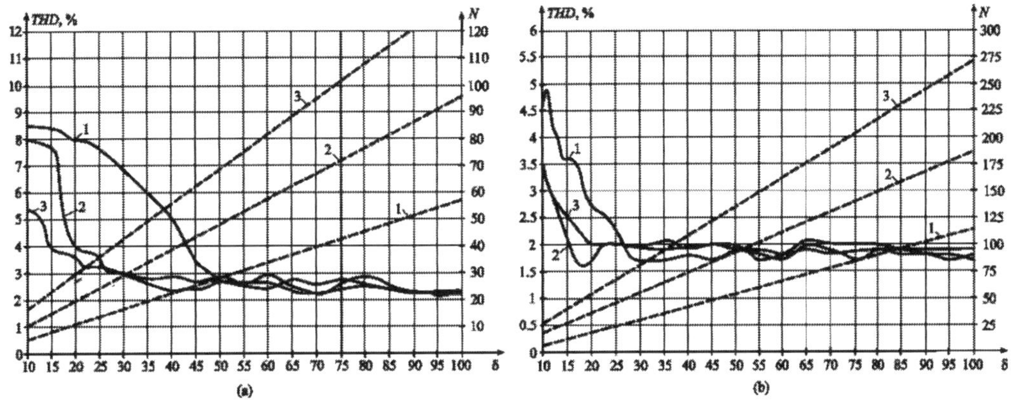

Fig. 3. THD (solid line) and N (dashed line) vs the damping ratio δ at K_{rip} = 0,2 (a) and K_{rip} = 0,1 (b) for different U^*: 1 – U^* = 0,8; 2 – U^* = 0,5; 3 – U^* = 0,2.

transient in every cycle of the generator. Visible reduction of THD with growth of N at the same time are observed from some minimal value of damping ratio δ_{min}. Moreover, as K_{rip} becomes smaller, δ_{min} becomes smaller too. The speed of decreasing THD , when $\delta > \delta_{min}$, depends on ripple rate and normalized amplitude of the output voltage. For example, when K_{rip} = 0,3 and U^* = 0,2 ($\delta_{min} \approx 16$), the speed of decreasing of THD is less than when U^* = 0,2 ($\delta_{min} \approx 14$) and U^* = 0,2 ($\delta_{min} \approx 23$). If damping ratio grows further, THD will come close to steady state value. This is seen especially when K_{rip} = 0,1. It can be explained by the following, when some value of N is reached, the farther increasing of the number of operating cycles in the half-period of the formed quasi-sinusoid wave does not lead to visible decreasing of THD. When K_{rip} = 0,4, THD does not reach steady state values in the observed range. It should be noted that when K_{rip} = 0,1 and K_{rip} = 0,2, in the range of changing δ THD is not higher than 12%. The mentioned characteristics allow to choose a near-optimal value of inductance for the current forming inductor which can provide relatively small dynamic losses in switches and the needed value of THD.

Obviously, the inductance of the current forming inductor should be as small as possible to reduce the calculated power of it. So, it is appropriate to make a formation of quasi-sinusoidal current when values of attenuation rate ($\delta = T/\tau = T{\cdot}R_{load}/L$) are respectively high. In this case, low values of THD are provided. However, the number of operating cycles of switches N with average frequency of transistors switching ($f_{S\,\mathrm{av}} = 2{\cdot}N{\cdot}f$) and level of dynamic losses can become unacceptably high. Therefore, the choice of the value of δ should be decided on the base of the complex parameters: power of current forming inductor, the value of THD, the average frequency of the transistors switching and the losses in them. Values of ripple factor recommended for practical using are commonly in range K_{rip} = 0,2÷0,3.

Table 1. The results of decomposition in normalized (to 1st harmonic) form.

Current	Number of harmonic									
	1	2	3	4	5	6	7	8	9	10
ASC, $\times 10^{-3}$	10^3	347	0	69	0	30	0	17	0	11
AQC[1], $\times 10^{-3}$	10^3	345	5.4	63	5.1	28	8.3	18	14	21
SQC[1], $\times 10^{-3}$	10^3	1.8	6.7	1.9	3.9	1.7	3.9	1.2	7.9	1.3
AQC[2], $\times 10^{-3}$	10^3	380	29	79	27	30	18	34	42	33
SQC[2], $\times 10^{-3}$	10^3	2.5	29.5	1.2	29.6	1.8	19.6	1	46	0.8
[1]K_{rip} = 0,2; [2]K_{rip} = 0,3										

To evaluate the influence of half-wave asymmetry on harmonic factor, spectra contents of asymmetrical sinusoidal current (ASC), asymmetrical quasi-sinusoidal current (AQC) and symmetrical quasi-sinusoidal current (SQC) were investigated.

Fourier series expansion of asymmetrical sinusoidal current on harmonics was made by known formulas, and analyzing of spectrums of AQC and SQC was made in PSpice with calculations of THD. The results of this decomposition in normalized (to 1st harmonic) form for $K_{rip} = 0{,}2$ and $K_{rip} = 0{,}3$ when $S = 10$, $\delta = 20$, are shown in Table 1 and Table 2.

Table 2. The results of decomposition in normalized (to 1st harmonic) form.

Current	Number of harmonic									
	11	12	13	14	15	16	17	18	19	20
ASC, $\times10^{-3}$	0	7.3	0	5.3	0	4.1	0	3.2	0	2.6
AQC[1], $\times10^{-3}$	13	6.8	19	25	17	12	26	28	21	9.9
SQC[1], $\times10^{-3}$	5	1.2	1.7	2.1	16	0.6	22	3	26	2.1
AQC[2], $\times10^{-3}$	25	21	19	29	12	44	69	33	13	32
SQC[2], $\times10^{-3}$	27	1	25	1.5	13.4	1.9	69	2	14	2.3
[1]$K_{rip} = 0{,}2$; [2]$K_{rip} = 0{,}3$										

Analyzing the data from tables, we can draw following results:

- The spectral composition of AQC is represented by even-numbered harmonics (the more significant are the 2nd, the 4th and the 6th).
- The value of the asymmetric current THD is defined by the 2nd harmonic, and its weight in the harmonic spectrum is in 5 and more times bigger than the other higher harmonics. The weight of the 2nd harmonics of ASC and AQC is approximately the same, and their THD has not much difference (with $S = 10$ and $K_{rip} = 0{,}2$ THD of ASC is 35,61%, and THD of AQC is 36,67%).
- The weight of the low-order high harmonics in AQC spectrum is far more than the weight of the same harmonics of SQC spectrum. So, the half-wave asymmetry takes the main part in setting the value of THD.
- With increasing of the ripple factor there is also an increasing of the high harmonics weight in the spectrum of the formed SQC. It is explained by the increasing of THD. For example, when $K_{rip} = 0{,}2$ the total harmonic distortion of SQC is 75% and when $K_{rip} = 0{,}3$ it is 11,7%.

4 Conclusions

The investigation of the asymmetric quasi-sinusoidal current harmonic composition and the symmetric quasi-sinusoidal current harmonic composition shows that the influence of half-wave asymmetry on the value of THD is much greater than the influence of non-ideality of symmetric quasi-sinusoidal current.

To provide THD of the symmetric quasi-sinusoidal current with value not above 12% and with the ripple factor equal to (20÷30)%, the period of the formed current need to be bigger not less than in 10 times of magnitude of the time constant of the current forming circuit with the minimal load resistance.

References

1. B.A. Baginskiy, V.V. Grebennikov, S.V. Obraztsov, *6th Int. Scientific and Practical Conf. of Students, Postgraduates and Young Scientists "Modern Techniques and Technology"*, 74 (2000)
2. B.A. Baginskii, V.V. Grebennikov, B.M. Nigof, D.N. Ogorodnikov, E.V. Yaroslavtsev, Instrum. Experim. Tech., **44(2)**, 243 (2001)

Multisensor transducer based on a parallel fiber optic digital-to-analog converter

Vladimir Grechishnikov[1], *Olga* Teryaeva[1,*], and *Vyacheslav* Arefiev[1]

[1] Samara National Research University, 443086 Samara, Russia

Abstract. Considered possibility of creating a multisensory information converter (MSPI) based on new fiber-optic functional element-digital-to-analog (DAC) fiber optic converter. The use of DAC fiber-optic provides jamming immunity combined with low weight and cost of indicators .Because of that MSPI scheme was developed based on parallel DAC fiber-optic (Russian Federation Patent 157416). We came up with an equation for parallel DAC fiber-optic. An eleborate general mathematical model of the proposed converter. Developed a method for reducing conversion errors by placing the DAC transfer function between i and i + 1 ADC quantization levels. By using this model it allows you to obtain reliable information about the technical capabilities of a converter without the need for costly experiments.

1 Introduction

Fiber-optic digital multisensor transducers (MSTs) are a class of multi-input binary signal transducers designed for collection of data on threshold parameters of the controlled object, their transmission over a shared channel and restoration at the output in a form suitable for control and monitoring systems application.

The main competitive advantage of transducers of this class compared with their electronic counterparts is evident in design of multivariable control systems of spatially distributed objects operating under strong electromagnetic fields, as well as in explosive and environmentally hazardous environments. Such operation conditions are typical for control systems used for space purposes, conventional and nuclear power engineering, chemical industry and other fields.

The main requirements of MSTs in these conditions are: immunity to electromagnetic interference, zero risk of abnormal spark formation, high number of monitored parameters and accuracy of reproduction of the input signal values, high performance — typical for electrically passive control systems with pneumatic and hydraulic actuating units.

The problems of development of MSTs and their components are expounded in patent literature and research work of Russian and foreign authors among them V.M. Grechishnikov, V.A. Zelensky, O.A. Kulish, G.I. Leonovich, J. H. Hong, C. M. Verber, I. A. Galton, Y-K. Chen, A. Leven whose work is highlighted [1, 2].

* Corresponding author: arefeva_olga@inbox.ru

© The Authors, published by EDP Sciences. This is an open access article distributed under the terms of the Creative Commons Attribution License 4.0 (http://creativecommons.org/licenses/by/4.0/).

2 Parallel FODAC-based MST

Examples of MST based on parallel digital to analog fiber optic converter are discussed in section [3, 4]. In the existing devices there are significant errors in generalization of quantized signal at the output of the DAC due to unevenness of optical radiation entering into the channels limiting metrological characteristics of the device as a whole In order to address this problem, the modified MST design was proposed [1]. The problem of non-uniformity of optical input in MST described above is solved by replacing fixed-value resistors in the LED supply circuit with trimming resistors, and supplying offset 0.5V of quantization increment of FODAC output parameter to the input of the amplifier. A diagram of MST is shown in Figure 1.

MST circuit includes a reference voltage source 1, a four-channel binary converter unit (BCU), a FODAC, a common fiber optic light guide 14, a photo detector 15, an amplifier 16, an ADC 17. Each channel of BCU includes an electronic key 2, a trimming resistor 3 and LED 4. FODAC includes transmitting optical fibers 5, transmitting spherical lenses 6, weighting elements 7-10, receiving spherical lenses 11, receiving optical fibers 12, a Y-coupler 13.

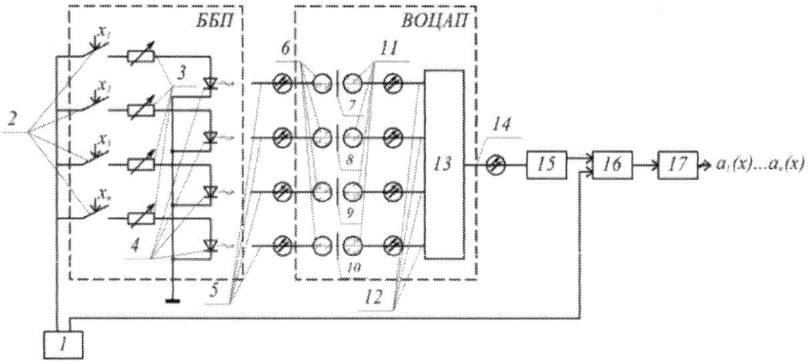

Fig. 1. Diagram of parallel FODAC-based MST.

MST operates as follows. When keys 2 are closed, REF 1 sets the level of direct optical radiation at the output of LED 4. Optical radiation flows through the transmitting optical fiber 5 and the transmitting spherical lens 6 to weighting elements 7-10. They provide weighting of each light flux in accordance with the law 2^{-i}. Then each of the light flows is transmitted to the Y-coupler through the receiving spherical lens 11 and the receiving optical fiber 12. Optical signals are generated and transmitted to the final position of the actuating mechanism. Optical signal from the output of the Y-coupler is transmitted through the common optical fiber 14 to the photodetector 15. Then the signal is amplified at the amplifier 16 and receives offset voltage U_{ref} from the second output of REF. Upon amplification and offsetting, the voltage is supplied to the input of ADC where it is converted into code $\{a_i\}$ that explicitly corresponds to terminal positions of actuators $\{x_i\}$.

3 Generalized mathematical model of a parallel FODAC-based MST

A mathematical model of a parallel FODAC was derived in [4, 5]. For ADC output signal to match DAC input signal, it is necessary for DAC quantization increment (Δ_q^{DAC}) to match the value of the ADC quantization increment (Δ_q^{ADC}): $\Delta_q^{ADC} = \Delta_q^{DAC}$. Given that

$$\Delta_q^{ADC} = \frac{U_{ref}}{2^n} \text{, and } \Delta_q^{DAC} = p_0 x_0 2^{-(n-1)} \text{,} \tag{1}$$

where U_{ref} is ADC reference voltage, n is ADC discharge, p_0 is step of quantization of the analog output. We obtain:

$$p_0 x_0 2^{-(n-1)} = \frac{U_{ref}}{2^n} \text{, } p_0 R_{feed} S_{PD} 2^{-(n-1)} = \frac{U_{ref}}{2^n} \text{.} \tag{2}$$

A group of expressions below follows from (2):

$$R_{feed} = \frac{U_{ref}}{2p_0 S_{PD}} \text{; } p_0 = \frac{U_{ref}}{2R_{feed} S_{PD}} \text{; } S_{PD} = \frac{U_{ref}}{2p_0 R_{feed}} \text{.} \tag{3}$$

The equality $\Delta_q^{ADC} = \Delta_q^{DAC}$ can be obtained by variation of feedback resistance, optical power level at the output of the BCU or changing of photodevice sensitivity.

As shown in Figure 2, it is evident that the scale of DAC transfer characteristic coincides with the levels of ADC quantization, and the slightest signal instability can lead to significant errors in ADC operation. To avoid these, it is necessary to shift U_{off} up by an amount equal to: $\Delta U_{off} = 0,5\Delta_q^{DAC}$ (parameter U^*_{DAC}). The transformed offset parameter has DAC output quantization levels between i and $i+1$ of ADC quantization levels (see Fig. 4). This provides one-to-one correspondence between the output code of ADC ($a_0, a_1, ..., a_{n-1}$) and the input code of the data collection device ($x_0, x_1, ..., x_{n-1}$) set by the position of the 2.

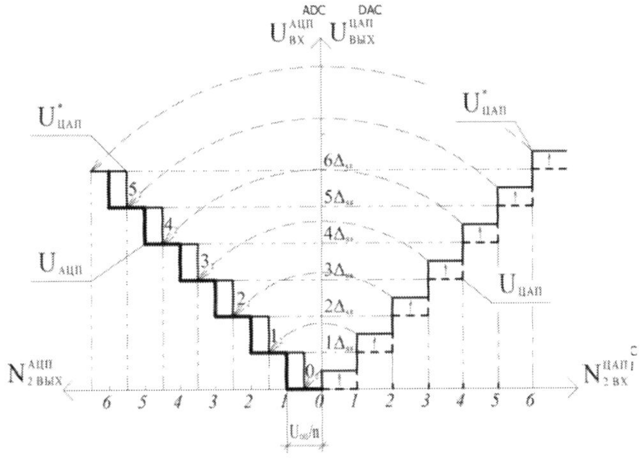

Fig. 2. Parameters of conversion provided by FODAC and ADC.

Through these operations, the following one-to-one correspondence is achieved: $\{x_i\} = \{a_i\}$

The signal at the output of the amplifier:

$$U_\Sigma(x) = S_{PD}\, R_{feed}\, \vartheta_i\, p_0\, N[\{x_i\}] + \Delta U_{off} \tag{4}$$

where p_0 is quantization interval of the analog output signal; ϑ_i is the coefficient of energy losses in fiber-optic communication lines; S_{PD} is the sensitivity of the detector; R_{feed} is feedback resistance of a transducer transforming photoelectric current into voltage; ΔU_{off} is zero offset voltage; $N[\{x_i\}] = P^*$.

Let us assume ADC to be voltage-to-code converter (VCC) with parallel conversion [6]. The output signal of i comparator of VCC unit can be set by a single function with the following form:

$$\mathrm{e}_m = \begin{cases} 1 \quad at \quad U_\Sigma(x) + \Delta \mathrm{U}_{\mathrm{off}} \geq \dfrac{m}{n} U_{\mathrm{ref}}; \\ 0 \quad \text{при} \quad U_\Sigma(x) + \Delta \mathrm{U}_{\mathrm{off}} < \dfrac{m}{n} U_{\mathrm{ref}}, \end{cases} \tag{5}$$

where U_{ref} is ADC reference voltage, $m = 2^{n-1}$.

Using Boolean algebra rules, we can obtain expressions for the values of bit digits of the output code as follows:

$$\begin{aligned} a_0(x) &= \overline{\overline{b_1(x)}\,\overline{b_3(x)}\,\overline{b_5(x)}\,\overline{b_7(x)}}; \\ a_1(x) &= \overline{\overline{b_2(x)}\,\overline{b_3(x)}\,\overline{b_6(x)}\,\overline{b_7(x)}}; \\ a_2(x) &= \overline{\overline{b_4(x)}\,\overline{b_5(x)}\,\overline{b_5(x)}\,\overline{b_7(x)}}, \end{aligned} \tag{6}$$

where $\begin{cases} b_i(x) = e_i(x)\overline{e_{i+1}(x)}; \\ b_m(x) = e_m(x)\overline{e_0(x)}, \end{cases}$ where $i = 1, \ldots, (m-1)$.

Thus, the resulting mathematical model makes it possible for us to investigate a set of metrological characteristics of optical DACs, taking into account deviations of design parameters of the transducer ($S_{\Phi\Pi}$, R_{OC}, ϑ_i, p_0 $N[\{x_i\}]$, ΔU_{CM}) from the nominal value, in accordance with the procedures set forth in [6].

4 Conclusion

We propose the original design of parallel FODAC-based MST for control systems with electrically passive pneumatic and/or hydraulic actuators. An equation for FODAC conversion was obtained, taking into account the passive nature of its optical circuit. We considered a method to reduce the error of conversion by placing DAC transfer factor between i and $i+1$ ADC quantization levels. A generalized mathematical model of the proposed data collection device was developed to display data transduction processes in its analog, analog-to-digital and digital functional elements. Application of this model makes it possible to obtain reliable information on the technical capabilities of the described transducer with no need for costly physical experiments.

References

1. S. Chin-Chong Tseng, US Patent 3985423 (1974)
2. V.M. Grechishnikov, O.V. Terjaeva, RF Patent 157416 (2015) (in Russian)

3. V.M. Grechishnikov, V.G. Domrachev, O.V. Teryaeva, A.A. Yudin, Measurement Techniques **57,** 1309 (2015)
4. V.M. Grechishnikov, O.V. Teryaeva, Russian Aeronautics **59**, 426 (2016)
5. V.M. Grechishnikov, V.G. Domrachev, I.V. Retinskaya, O.V. Teryaeva, Measurement Techniques **58**, 1127 (2016)
6. V.M. Grechishnikov, *Metrologiya i radioizmereniya* (Publishing House of Samara National Research University, 2007) (in Russian)

Personal navigation system based on MEMS gyroscope

Lo Van Hao[1,*], and *Tamara* Nesterenko[1]

[1]National Research Tomsk Polytechnic University, 634050 Tomsk, Russia

Abstract. The paper presents a configuration of the modern personal navigation system and a new method of calculation of the distance traveled. The proposed navigation system is used to estimate a person's position and consists of three micromechanical gyroscopes: two at the knee and one at the waist. The knee attached gyroscope gives the distance traveled information, and the waist attached gyroscope gives the heading information. During the leg movement, the angular displacements of knees repeat at each period (each step of the walking person). By summing up all the stride lengths, the distance that a person has walked can be determined.

1 Introduction

In the field of space research, civil and military technology, navigation systems are being intensively developed for and applied to solving problems of orientation and navigation of mobile devices [1-2]. Currently, navigation systems are extensively utilized in the field of personal navigation. There is a variety of the implementation techniques for the personal navigation system. One of the most popular is the Global Positioning System (GPS) which, however, does not always provide the access to indoor operations and restrained urban conditions. Therefore, to track a person, considerable attention should be paid to a self-contained pedestrian navigation system with inertial sensors, *i.e.* inertial navigation system.

In typical inertial navigation systems, the source information is measured with the angular and translational movement sensors (gyroscopic devices and accelerometers) [3-5]. This information is processed by a computer module with a view to obtain such navigation parameters of mobile objects as walking speed, orientation, and position. In order to minimize the size and the number of sensors, the authors present a personal miniature inertial navigation system consisting of three micromechanical gyroscopes.

2 Circuit configuration of personal navigation system

The personal navigation system presented in this work consists of three uniaxial micro-gyroscopes of the LL-type in which the inertial mass executes a plane-parallel motion. Micromechanical gyroscopes comprise the electromechanical and circuit modules. Its operating principle is based on the energy transfer at an angular rate of an object between

* Corresponding author: lovanhao.pro@gmail.com

© The Authors, published by EDP Sciences. This is an open access article distributed under the terms of the Creative Commons Attribution License 4.0 (http://creativecommons.org/licenses/by/4.0/).

the two perpendicular directions (primary and secondary oscillations) [6]. Sensor converts the angular rates of an object into changes of capacitances transformed by the circuit module, based on lock-in technique [7-8], into the output signals proportional to the measured rates.

Figure 1 presents a generalized flow chart of the personal navigation system. Radio channel transmits and receives the personal navigation signals which are displayed on the monitor. MMG measure the angular rate within the coordinate system of the device, and the object rotation relative to its original position can be calculated using the digital integration of the gyroscope time readings and the time interval between the recordings. The personal navigation system operates as follows. Gyroscopes measure the angular rates and convert them into digital signals which then transfer to the computer module. The obtained distance is transferred from the computer module to the computer *via* the radio module which incorporates the LabVIEW [9] system-design platform to display the signal.

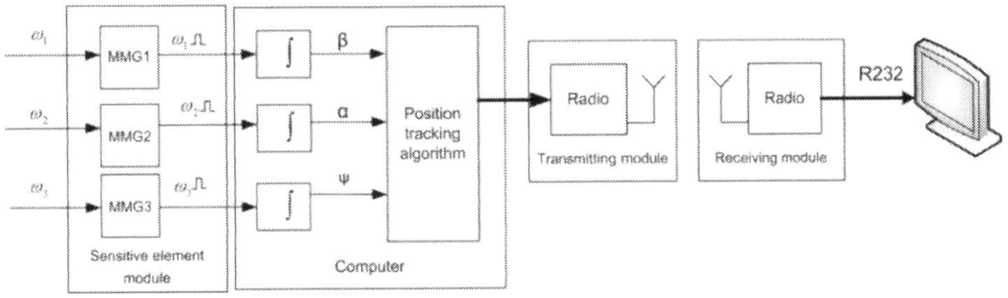

Fig.1. Generalized flow chart of personal navigation system.

3 Distance calculation method

The proposed system of personal navigation is based on MMG which measure the angular rate of walking speed. Here, we propose to use three micromechanical gyroscopes, two attached near knees and one attached to waist as illustrated in Figure 2. Angular rates of the leg movement are measured by the two knee attached gyroscopes. The angular rate of the upper body swing is measured by the waist attached gyroscope.

From Figure 2, two knee gyroscopes are attached such that the sensitive axis should be parallel to *OY*-axis. Let us consider the first step of a person. The left leg is a lever which turns around *O*1 reference point by β angle. The angular rate of the left leg is measured by the knee attached gyroscope *1*. The right leg is a lever which turns around *O2* reference point by α angle. The angular rate of the right leg is measured by the knee attached gyroscope *2*. And the angular rate of a person around *Z*-axis is measured by the waist attached gyroscope *3* and gives us the angle of person's rotation.

However, the MMG output signal contains the error which accumulates as long as there is integration. To bound the error, the MMG should be calibrated and the data should be filtrated. Moreover, a new method, called Zero Angular Displacement Update (ZADU), is introduced in this work.

The key idea of this method is integration provided separately at each period. The first step is carried out during T1 period and measured by gyroscopes *1* and *2*. As can be seen from Figure 2, the right leg carries out not only the angular but also translational movement.

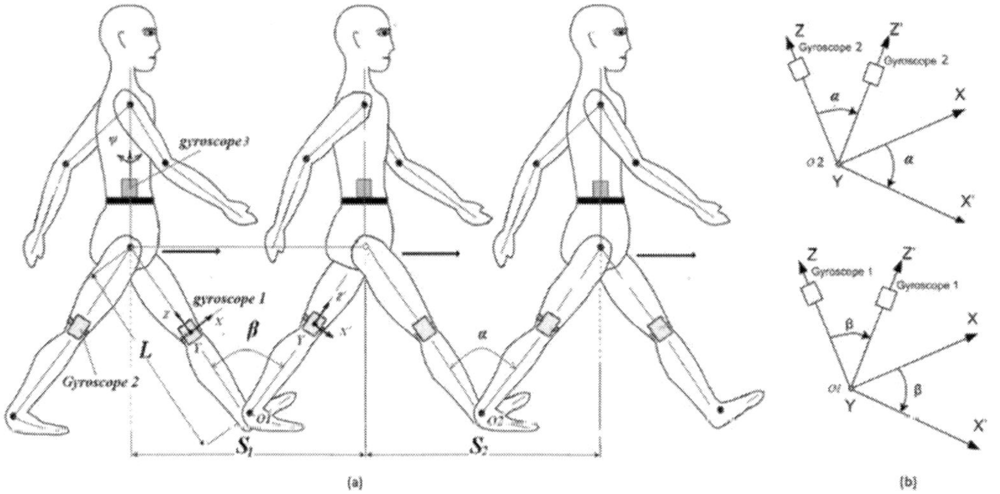

Fig.2. Sensor deployments: a – walking person; b – coordinate system.

The translatory acceleration has a strong effect on the MMG accuracy in its sensor mode. Therefore, the movement information is not given by gyroscope *2* attached to the right leg. The left leg being the lever *L*, carries out the step by the displacement angle β which is calculated by the integration of the angular rate obtained by gyroscope *1*. During the next period of T2, the second step is carried out. In this case, the signal from gyroscope 2 is not received because of the effect of the translatory acceleration, so gyroscope 2 should be used. The displacement angle α produced by the right leg is calculated from integration of the angular rate provided by gyroscope *2*.

Figure 3 demonstrates signals measured by gyroscopes *1* and *2*.

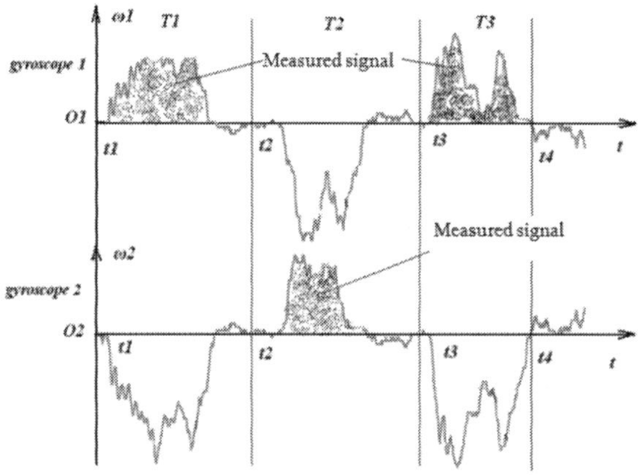

Fig.3. Signals measured by gyroscopes.

The data integration obtained by the gyroscopes is performed in series. During the first period, the data from gyroscope *1* are integrated and recorded for the calculation of the angular displacement. And after this period of time, the MMG angular displacement value is reset back to zero. During the second period, the data from gyroscope *2* are integrated and recorded for the calculation of the angular displacement. Similar to the first period of time, after this period the MMG angular displacement value is also reset back to zero. Thus, in ZADU method the initial value of each time integral is always zero, and the next calculation does not include the errors from the last calculation. So the output navigation signals contain the minimum error.

Angles β and α can be measured by discrete integration of the angular rate. The heading information is obtained *via* gyroscope *3*.

$$\beta=\beta_0+\int_{t1}^{t2}\omega_1\times dt\ ;\ \alpha=\alpha_0+\int_{t2}^{t3}\omega_2\cdot dt\ ;\ \psi=\psi_0+\int_{t1}^{tn}\omega_3\cdot dt, \tag{1}$$

where ω_1, ω_2 are the angular rates of human legs measured by knee attached gyroscopes *1* and *2*; ω_3 is the angular rate of upper body swing measured by the waist attached gyroscope *3*.

Thus, the following equation is obtained for the stride length *S*:

$$S=S_1+S_2=L\cdot\left[\sqrt{2\cdot(1-\cos\beta)}+\sqrt{2\cdot(1-\cos\alpha)}\right], \tag{2}$$

where *L* is the leg length; *S*1 is the stride length of the first period; *S*2 is the stride length of the second period as shown in Figure 3.

By summing up all the stride lengths, the distance that a person has walked can be determined. Let *N* be the total number of strides, the distance covered then can be written as

$$\sum S=\sum_{i=1}^{N}L\cdot\left[\sqrt{2\cdot(1-\cos\beta_i)}+\sqrt{2\cdot(1-\cos\alpha_i)}\right] \tag{3}$$

4 Conclusions

The flow chart and the algorithm of the personal navigation system based on three gyroscopes were discussed. The proposed navigation system has the advantage of simplicity, small size, low cost and self-contained sensors. However, certain efforts should be provided to keep the angular displacements errors bounded. For this purpose, the Kalman filter and calibration of high-precision MMG were employed. In view of the absence of the feedback, most of MMG are characterized by unsteady operation under the dynamic loads. Therefore, the gyroscopes of compensation type used as inertial sensors will be the most promising.

References

1. C. Acar, A.M. Shkel, *MEMS Vibratory Gyroscopes Structural Approaches to Improve Robustness* (Springer, New York, 2008)
2. E.V. Golodnykh, V.N. Borikov, *SIBCON*, 6693635 (2013)
3. T. Jiang, et. al., Sens. Actuators A: Phys **128** (2006)
4. X. Liu, et. al., Sens. Actuators A: Phys **154** (2009)
5. V.J. Skorykh, A.N. Gormakov, I.A. Plotnikov, *KORUS*, 42 (2001)
6. T. Nesterenko, S. Vtorushin, E. Barbin, A. Koleda, IOP Conf. Ser.: Mater. Sci. Eng. **81** (2015)
7. P. Baranov, T. Nesterenko, E. Tsimbalist, E. Barbin, A. Koleda, S. Vtorushin, *IMEKO: Proc. 14th IMEKO TC10 Workshop on Tech.l Diag.* (2016)
8. P.F. Baranov, E.I. Tsimbalist, V.N. Borikov, D.G. Soltanova, *IMEKO*, 116100 (2015)
9. V.N. Borikov, P.F. Baranov, A.D. Bezshlyakh, *SIBCON*, 5044870 (2009)

Physico-mathematical modeling methods for the pressure distribution determination in the gas-dynamic bearing gap of the ball gyroscope

Anastasiya Ignatovskaya[1,*], and *Alexey* Golikov[1]

[1]Tomsk Polytechnic University, 634050 Tomsk, Russia

Abstract. The considering problem in the article connects with solution methods of the specific questions in gas-dynamic lubrication theory. The comparative analysis of analytical and numerical methods for the calculating gas-dynamic bearing characteristics is provided. The main applying aspects of them for the solving of specific gas lubrication theory points are presented. This research is carried out for the investigated gap bearing geometry for the designed ball gyroscope construction. The main mathematical equation and results of developed numerical simulation for the pressure distribution determination are shown.

1 Introduction

During the gyroscopic technology development, rotor bearing construction has remained one of the main gyroscopy problems. Gas lubrication application for the gyroscopic devices was determined by the tendency of increasing the gyroscope accuracy, rigidity and operating life. Using gas lubrication allowed decreasing frictional losses, enhance the durability and reliability. Gas bearings can work under the severe mechanical influences, increased radiation conditions, and in the wide range of temperatures and ambient pressure without losing its performance characteristics [1].

Ball gyroscopes have the specific field of application. Generally, they are used as sensitive elements for precision inertial navigation systems and gyro-stabilized platforms. The ball gyroscope construction is different from others in quite high resistance to overloads and impacts of external dynamic effects. The main type of gas bearings as a part of such gyroscopic devices are gas-dynamic bearings (GDB). This is due to a strong requirement for gyroscope weight and size parameters; gas-dynamic bearings do not require additional excess pressure sources such as gas bearings with an external pressurization.

The final selection of the rotor bearing type and parameters depends on the field of application, design features, and developing purposes. These aspects of research work were considered in the previous article [2]. It should be emphasized that hemispherical configuration of the GDB working surfaces has been selected as of providing the sufficient reserves for the load capacity and stiffness, as well as the spin axis position stability and the low level of natural vibration frequency.

* Corresponding author: nastena@tpu.ru

© The Authors, published by EDP Sciences. This is an open access article distributed under the terms of the Creative Commons Attribution License 4.0 (http://creativecommons.org/licenses/by/4.0/).

It is expected that basing on such type of gyroscope with the hemispherical configuration of GDB will allow creating a reliable and sufficiently accurate inertial navigation system for the solution of relevant problems of metrological drilling process maintaining.

2 Materials and methods

Gas lubrication theory is a part of viscous fluid dynamics where the fact that the gas layer thickness is much less than its other dimensions. The gas lubrication flow is generally laminar, and inertial forces are negligible in comparison with viscous friction forces. Therefore, lubrication theory basic equation (Reynolds equation) is a consequence of the equilibrium equations between forces of viscous friction and normal pressure [3]. Basically, the Reynolds equation solution is a key component of the solving mathematical modeling problems.

Reynolds differential equation for the gas lubrication is [4, 5]:

$$\frac{\partial}{\partial x}\left(h^3 \frac{p}{\mu}\frac{\partial p}{\partial x}\right)+\frac{\partial}{\partial z}\left(h^3 \frac{p}{\mu}\frac{\partial p}{\partial z}\right)=12pV_y+6\frac{\partial}{\partial x}(pV_x h)+6\frac{\partial}{\partial z}(pV_z h)+12h\frac{dp}{dt}, \quad (1)$$

where V_x, V_y, V_z - relative velocity components of the working surfaces in the tangential, radial and axial direction, respectively; h - lubricating layer thickness in the considering section; μ - dynamic viscosity; p - pressure. Taking into account the injection ability of grooves (if available) and geometric features of the bearing profile, Reynolds equation becomes in a more complicated form.

The main calculating methods of the gas bearing characteristics are analytical and numerical methods; among the lasts - application software packages. The most important principle components of any mathematical modeling methods are [4]:

1. Preparation of the phenomenon, process or task mathematical model;
2. The transition from the continuous mathematical model to the discrete model;
3. Writing the algorithm for the discrete model (sequence of computational steps, and transferring to a computer system);
4. Software package configuration for the certain task;
5. Getting results and their analysis.

Certainly, the analytical methods are the most preferable methods of mathematical modeling. They provide the deeper dive into the physical nature occurring in the lubricating layer phenomenon and, therefore, help to avoid mistakes in the statement of task. Additionally, using the special application oriented software packages for solving the system of equations which was obtained by analytical method, it is possible to carry out the high-precision calculation of "rotor-bearing" system for evaluating effectiveness, reliability and efficiency of the designed unit.

On the other hand, numerical methods over the past few years have also gained the quite powerful tools for the certain gas dynamics problems solving. They can solve the problems of three-dimensional steady and unsteady flows, laminar and turbulent flows with free surfaces, multiphase flow problems, with and without cavitations, chemical reactions, and etc. Application software packages are used for simplifying and advancing the mathematical modeling. But it should be realized that it occurs with some restrictions.

Despite of various advantages and disadvantages of a particular method, the theoretical and numerical simulation of the gas bearing characteristics exceptionally large and significant. Whereas, the experiment in this area is extremely difficult and expensive as of ultrathin gap (several microns) between the working surfaces.

The following part, numerical experiment for the pressure distribution determination in the gas-dynamic bearing gap of the developed ball gyroscope using ANSYS software will be presented.

3 Numerical simulation

The main gas-dynamic bearing characteristics are: pressure distribution in the gap bearing, bearing capacity accordingly, bearing rigidity, and the value of viscous friction momentum. Geometry of interest for the ball rotor with gas-dynamic suspension is represented in Figure 1 schematically.

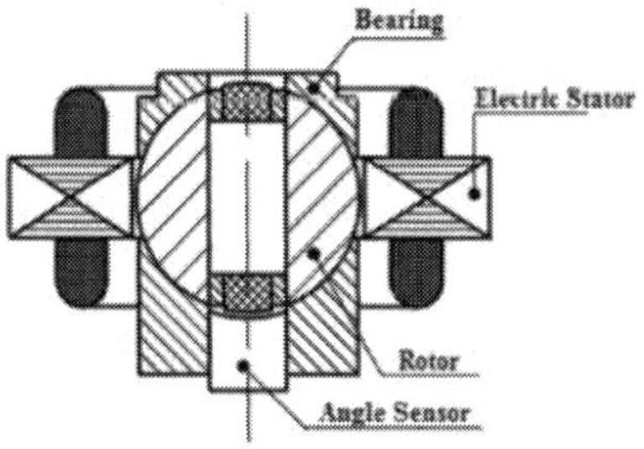

Fig. 1. Considering gap geometry of the "rotor-bearing" system.

The rotor is a ball with diameter 28,587 mm. It is made from steel SHKH15. During the rotation it is supported by two hemispherical gas-dynamic bearings. The hemispherical bowls surface is smooth and modified so that the spherical diameter for 10 microns more than the actual rotor diameter (5 mm per side). This difference provides an initial clearance which is obligatory for the working at the gas lubrication operational mode. The ball rotor is driven by the magnetic field of the three-phase asynchronous electric stator. Maximum voltage for the stator is 40 V and frequency is 1000 Hz. The rotor has three rotational degrees of freedom. In case, when the rotor and stator magnetic field rotating axes are coincided, the electromagnetic torque is applied to the rotor. The nominal rotor speed is 16 000 rev/min; the maximum angular momentum of the ball gyroscope is 0,013 N·m. The total layout length is 68 mm and the diameter - 77 mm; weight - not more than 1 kg. Dimensions are not the final [6].

The results of the gas-dynamic bearing calculation modeling for determination of the pressure distribution in the gap are provided in the following. The calculations were performed using application software package module in ANSYS - FLUENT. The mathematical formulation of this problem, in this case, has the general terms, as package flexibility requires from developers to cover the widest range of tasks. Besides of the aforementioned geometrical parameters, it is necessary to specify that the air was selected as the gaseous medium; type of gaseous medium model - incompressible ideal gas; conditions for pressure and temperature - normal. The mathematical model is based on solving the system of equations of fundamental laws of mass, momentum, and energy conservation. The system closes the initial and boundary conditions, as well as defining relations [6].

The pressure distribution across the gap on the ball rotor and hemispherical bowls surfaces are presented in Figure 2 *a)* and *b)*.

According to numerical simulation, the total pressure has the value up to 2067,3 Pa. The pressure distribution models on the bearings provided as spheres as rotor geometry. It is formed by two hemispherical bowls. This form is quite appropriate since the problem is axially symmetric. The models show that the maximal pressure is located in the equator areas as the radius here is maximal. Also, the width of the equatorial maximal pressure band on the rotor is greater than the hemispherical bowls. This is due to the fact that the rotor is movable part relative to the fixed bearing system. So, in the character of pressure distribution, dynamic component of the pressure plays an important role from the rotor rotation. Since the more the rotation speed of the moving part, the more contribution dynamic component, and it makes the total operating pressure higher.

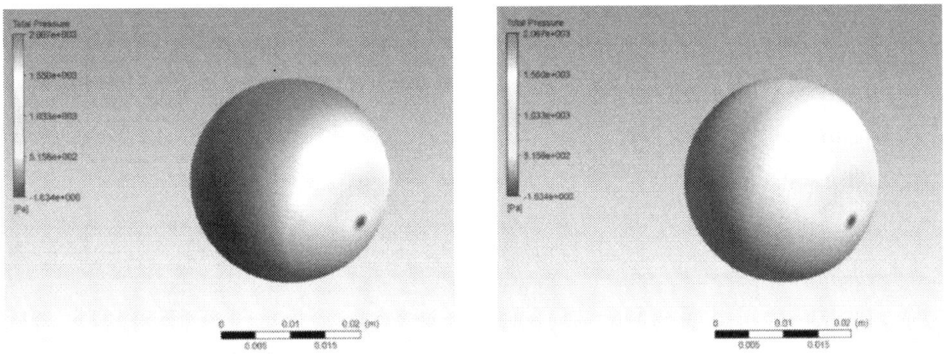

a) *b)*

Fig. 2. *a)* 3D model of the pressure distribution on the ball rotor surface at the nominal rotation speed 16 000 rev/min; *b)* 3D model of the pressure distribution on the hemispherical bowl surface at the same nominal rotation speed.

Using the given results about the pressure distribution field, it is possible to calculate other characteristics as following: load bearing capacity, rigidity, and provide estimation of viscous friction momentum. These calculations are carried out by analytical way and presented in the [6].

4 Conclusions

Currently, the numerical methods which are used for solving the gas lubrication problems cannot be as a full and enough replacement of analytical methods. Modern software packages are very objective. The complete solution of the gas lubrication mixed problem including calculation of the gas flow in the lubrication layer, temperature field determination, solution of dynamics and optimization problems cannot be carried out in any of the existing special software packages. For the solution of certain and application tasks, for instance, determination of the pressure field in the bearing gap, as was considered in the article, numerical methods can be like an accompanying tool. By their means, the preliminary assessment of the gas bearing parameters can be achieved. Also, some basic processes can be visualized, as well as using the program results for further calculations.

Recognizing the versatility and powerful of numerical methods, the analytical methods should not be underestimated. Based on their advantages, it is possible to analyze some specific and limit cases which allow checking the solution correctness obtained by numerical methods.

References

1. D.B. Smirnov, *Chislennoe modelirovanie zadach gazovoj smazki na osnove uravnenija Rejnoldsa* (Polytech, Saint-Petersburg, 2008) (in Russian)
2. A.A. Ignatovskaya, A.N. Golikov, J. Phys.: Conf. Ser. **671**, 012020 (2016)

3. I.E. Sipenkov, A.Yu. Filippov, Yu.Ya. Boldyrev, B.S. Grigoriev, N.D. Zablotchkiy, G.A. Luchin, T. V. Panich, *Precizionnye gazovye podshipniki* (Electropribor, St-Petesburg, 2007) (in Russian)
4. N.S. Grassam, J.W. Powell, *Podshipniki s gazovoj smazkoj* (Mir, Moscow, 1964) (Translation from English)
5. A.V. Dubinin, K.V. Smolyan, Prib. **S5**, 169 (2012)
6. A.A. Ignatovskaya, A.N. Golikov, Matec Web Conf. **48**, 01003 (2016)

Modification of polymeric materials for 3D printing of external panels of nanosatellites

Dariya Isaeva[1], *Fedor* Simankin[1,*], *Yuriy* Doncov[1], and *Arkadiy Simankin*[2]

[1]National Research Tomsk Polytechnic University, 634050 Tomsk, Russia
[2]Tula State University, 300012 Tula, Russia

Abstract. The results of mechanical testing of plastic samples, produced by injection molding and 3D printing are shown. Strength properties of filled and non-filled polymers are compared. The applicability of 3D printing technology with filled polymer materials of external panels is evaluated.

1 Introduction

One of the active developing areas of spacecraft engineering [1-2] attracting large investments is the creation of miniature satellites or nanosatellites (spacecraft weighing 1-10 kg) including the so-called "student" satellites. These satellites as usual are made base on 'CubeSat' standard. This standard was developed by the researchers from California Polytechnic and Stanford universities. It is special standard for small cheap satellites, which are suitable for academic and educational purposes. Using of this standard gives the opportunity to place unique products in the same type of launchers containers thereby increasing the payload of rocket. However, despite significant levels of investment in the new industry all satellites are objects of single-unit production. Each satellite has unique design and technological solutions.

One of the modern directions of production development is advanced materials and technologies using. For example, in 2016 nanosatellite 'TPU-120' was sent to the International Space Station. External panels of this device was made of polymeric material using additive technology [3]. 'Honda Motor' together with 'Kabuki' (Japanese company specializing in additive technologies) has announced the release of a mini-electric vehicle 'Micro Commuter' with freestanding elements produced by 3D printer [4].

2 Materials properties comparison

Polymeric materials are increasingly used to create basic items for different purposes. This is because the plastic parts are much lighter analogues made of aluminium alloys. At the same time, the strength of the modified polymeric material is often the same as the metal analogues. A serious drawback of polymeric products is the need of using expensive and technologically complex equipment. In the case of a single-unit production, the price of the

* Corresponding author: simankinfa@tpu.ru

© The Authors, published by EDP Sciences. This is an open access article distributed under the terms of the Creative Commons Attribution License 4.0 (http://creativecommons.org/licenses/by/4.0/).

product significantly increases. Additive technologies allow to create parts of a very complex shape without excessive rigging. The cost of the product obtained in this way is already much lower.

However, this technology has drawbacks. The products have lower strength characteristics [5-6]. This is due to technology, since the seal material is applied in layers. Firstly, "solid-state" virtual product model is divided into layers with simple thickness in a special program 'Slicer'. The layers are formed while moving in the working plane of the Hot-End nozzle, which injects molten polymer. The nozzle moves in the required manner to the coordinates X and Y and forms a layer of any degree of complexity. Then there is a shift of the head in the Z-axis direction by an amount equals to the thickness of the layer. The resulting product has the property of anisotropy. The maximum strength should be observed only in the direction corresponding to the axis of the strand (machine direction), extruded from the head (Figure 1, a). The next strand is laid parallel after some time. Thus, strength in the transverse direction is provided only by the strands sticking surfaces. Wherein the adhesion of adjacent strands depends on temperature and is heterogeneous in length. For example, on the rod with length L and width B there are some characteristic sections I, II and III (Figure 1, b). Sections II and III have more strength than section I since the temperature of adjacent strands is high in this area and the strands clump together in a very durable material. After a certain threshold the surface temperature is insufficient for reliable adherence and strength for the rest of the plot strands is not high enough ((Figure 1, c). In the Z-axis direction, material has the lowest strength, especially if the area of the layer is large and the solidification of the material starts at high temperature. In this case, the surface of the strands is too cold to allow reliable of adhesion between adjacent strands. There is the need for additional heating of the surface of the printed layer before applying the next. In addition, under certain print modes the diameter of the strands is insufficient for fulfilling voids, i.e., material forms with stretched longitudinal cavities. Such material should be considered as orthotropic composite plastic, and the simulation of its properties can be carried out according to [8-11].

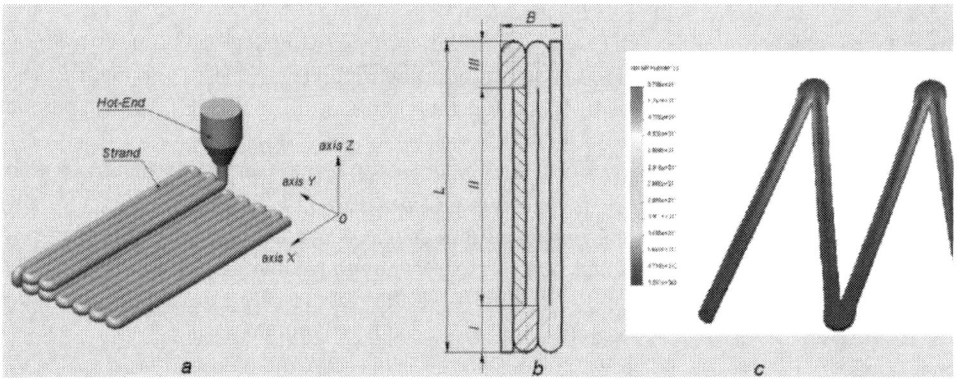

Fig. 1. Anisotropy of mechanical properties of material obtained by 3D printing.

The external panels of nanosatellites are designed to operate in very severe conditions of outer space. They should ensure maximum protection of the filling of the satellite. One of the perspective directions is the material reinforcement by carbon fiber. Reinforcement can be done with short fibers (filling) or a continuous carbon fiber. The axis of continuous fiber ccorresponds the axis of the extruded strand. Work on the creation of 3D printers enabling the printing of reinforced plastic is known [12]. However, printing with continuous reinforcement requires the creation of a fundamentally new type of equipment.

Despite the fact that continuous reinforcement does not solve the fundamental problem with the interlayer adhesion, the mechanical strength in the main direction increases significantly, making this method of obtaining body parts for nanosatellites very attractive. Increasing the strength in other areas can be provided by technology or software – for example, a partial shift of the elements of the layer on the Z-axis (software method implemented by the program slicer) or short-term local heating or cooling sections of the created layer (the technological method, requires the installation of additional equipment on the printer head).

3 Tests

The mechanical testing for evaluation of appropriate method of polymer modification was carried out. The samples were made from ABS-M30 plastic. Three different technologies had used for production of simples –injection molding (Figure 2, a) in the pressing chamber and the two methods of 3D printing. Samples of the resulting 3D printing have a structure formed by stacking of the strands in different ways along the direction of tensile load application (longitudinal strands) (Figure 2, b) and transversely to the direction of loading (transverse-strands) (Figure, 2 c). The samples were subjected to tension at a speed of 5 mm/min until complete destruction on the testing machine Gotech AI-7000M. The test results are shown in the chart (Figure, 3). Curve 1 corresponds to the average strength of samples obtained by injection molding. Curve 2 – the specimens printed on a 3D printer with longitudinal strands, curve 3 – the specimens printed on a 3d printer with the transverse strands

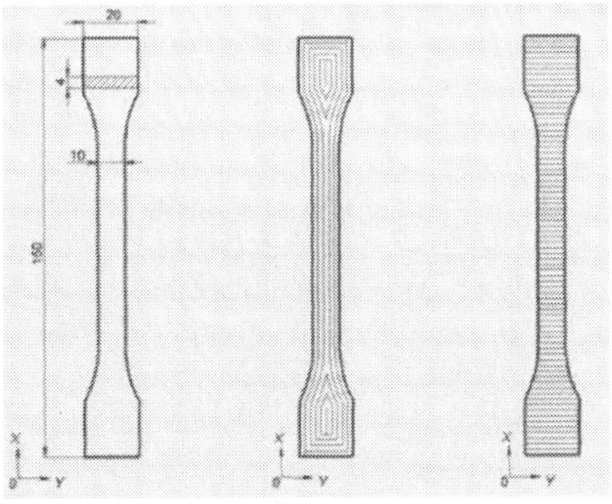

Fig. 2. Samples for mechanical testing.

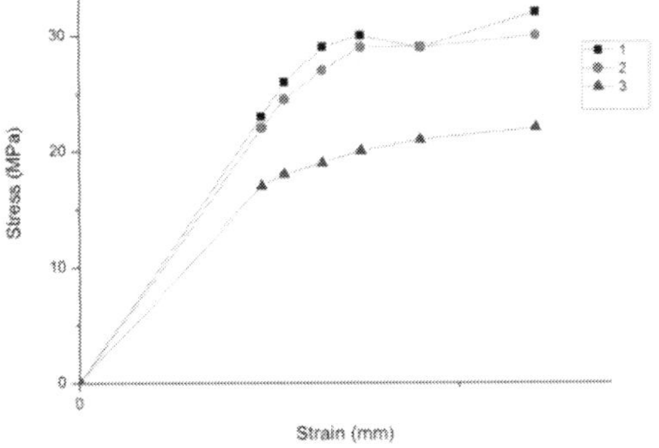

Fig. 3. Diagram of stress-strain in tension for the samples obtained in various ways: 1 –injection molding 2 – 3d print with the direction of the strands parallel to the applied load (longitudinal), 3 – 3d printing with the direction of the strands perpendicular to the applied load (transverse).

The graph shows that the samples obtained by injection molding and printed samples with longitudinal strands (along the X-axis) differ slightly (less than 5%). This drop in strength is related with the reduction in cross sectional area due to the presence in the samples stretched longitudinal cavities. Specimens with transverse arrangement of the strands (along the Y axis) have a lower strength (more than 20%). Thus, we can conclude that in this case the strength of the sample was determined by the strength of sticking of strands.

Comparison of mechanical characteristics of pure and composite (filled) materials [13] shows that the filled material has greater value of tensile strength and, as mentioned earlier, is characterized by a high degree of isotropy of mechanical properties. However, filled materials have the property of brittle fracture. That is not an advantage for device body. In the event of impact, such detail is rather split up to complete destruction, while the detail of the non-filled polymer has only permanent plastic deformation.

The strength of the printed composite sample, printed from the combination of a continuous fibers and plastic material as a binder is not possible to estimate because of open data technologies absence. The assumption that the strength of such sample is determined by the strength of the reinforcing fibers can be made in this case. The strength of the common carbon fiber materials now comes to 4000 MPa. In this case, product can have strength at least several times higher than the product of the non-filled plastic. The polymeric binder provides the printed product with elastic properties, allowing minimizing the effects of impact on the object. In some cases, it may be applied to the simultaneous reinforcement of short fibers and continuous central fibers.

4 Conclusions

Tough requirements to products used in outer space can be met using composite material reinforcement by carbon fibres. Performed mechanical testing of samples obtained by different methods (injection molding and 3D printing) show that the transverse arrangement of the material strands decreases strength characteristics to 20%. At the same time, strength increases in the reinforcement of the polymeric material with carbon fibres. This allows making a conclusion about the applicability of additive technologies for the manufacture of external panels for nanosatellites of polymeric material with central reinforcement strands of continuous carbon fiber.

References

1. K. Mertins, V. Ivanova, N. Natalinova, M. Alexandrova, MATEC Web of Conf. **48**, 06002, (2016)
2. K. Mertins, V. Ivanova, MATEC Web of Conf. **37**, 01025 (2015)
3. *Tomsk-TPU-120, the first 3D printed CubeSat Mission*, URL : https://directory.eoportal.org/web/eoportal/satellite-missions/t/tomsk-tpu-120
4. *Honda and Kabuku Inc.*, URL : http://www.conceptcarz.com/z26617/Honda-Micro-Commuter.aspx
5. B.A. Lyuckshin, *Computer simulation and design of composites* (Pub. house of the Sib.branch of the RAof S., 2015)
6. V. Dmitriev, Y. Britova, MATEC Web of Conf. **48**, 01005 (2016)
7. Yu. Britova, V. Dmitriev, T. Kostyuchenko, IOP Conference Series: Materials Science and Engineering **132(1)**, 012011 (2016)

8. N.A. Kupriyanov, F.A. Simankin and K.K. Manabaev IOP Conference Series: Materials Science and Engineering **124**, 012100 (2016)
9. A.A. Svetashkov, N. A Kupriyanov and K. K Manabaev, Fizicheskaya mezomechanika **18(6)**, 57 (2015)
10. A.A. Svetashkov, N.A Kupriyanov and K.K Manabaev, 2016 Key Engineering Materials **685**, 206 (2015)
11. F.A. Simankin, G.R. Ziyakaev, I.V. Kuznetsov, R.E. Lushnikov, IOP Conference Series: Materials Science and Engineering **124(1)**, 012072 (2016)
12. Xiaoyong Tian, Tengfei Liu, Chuncheng Yang, Qingrui Wang, Dichen Li, Composites: Part A **88** (2016)
13. *Metody ispytanij polimernyh materialov OOO «KOMEF»,* URL : http://www.komef.ru/metodisppolimer.pdf (in Russian)

MATEC Web of Conferences **102**, 01023 (2017) DOI: 10.1051/matecconf/201710201023
V International Forum for Young Scientists "Space Engineering"

Gyropendulum and vibration forces – experimental research

Veronica Ivanova[1,*], *Argen* Keremkulov[1], and *Nataliya* Kurkan[1]

[1]National Research Tomsk Polytechnic University, 634050 Tomsk, Russia

Abstract. It is well-known that gyroscope base vibration as usual has a bad influence on accuracy of gyroscopic instruments and systems. During running and testing gyroscopic instruments are exposed to translational vibration, the accuracy and reliability are reducing. With rising demands on the accuracy and reliability of the gyropendulum systems, the problem of finding effects of vibration methods of research base and the possible ways to fix it remains relevant. The purpose of this paper is to present the results of experimental research of vibration influence on the gyropendulum, by using new term "vibration force". Two main regimes were tested: upwards and down displaced centre of mass of gyropendulum.

1 Introduction

During running gyroscopic instruments are exposed to various disturbances that reduce the accuracy of gyros and can lead to missing of the device efficiency. Angular and translational vibration, vibro-impact and disturbance caused by linear accelerations are mechanical perturbations. Vibration is generated by external and internal sources.

By now, a large number of private tasks of the influence of vibrational perturbations on the gyroscopic devices had been resolved.

The level of external vibration can have frequency from 2 to 10000 Hz, the amplitude up to 1 mm, and vibration overload up to several dozens of g.

Most of all gyroscopic instruments and systems built on the basis of gyropendulums, i.e. the gyroscopes, the center of mass which mismatches with the center of intersection of the axes of suspension are exposed to translational vibration of the base.

2 Methods and materials

2.1 Theoretical background

The truncated differential equations of motion of gyro pendulum on a vibrating base with down displaced center of mass with the use of the concept of "vibrational force" can be written as [1-7]:

* Corresponding author: kurator@tpu.ru

© The Authors, published by EDP Sciences. This is an open access article distributed under the terms of the Creative Commons Attribution License 4.0 (http://creativecommons.org/licenses/by/4.0/).

$$\left.\begin{aligned}\dot{\psi}+\left(\frac{m\cdot g\cdot l}{H}+\frac{m\cdot(S_0\cdot\omega)^2}{2\cdot H}\right)\cdot\vartheta=0\\ -\dot{\vartheta}+\left(\frac{m\cdot g\cdot l}{H}+\frac{m\cdot(S_0\cdot\omega)^2}{2\cdot H}\right)\cdot\psi=0\end{aligned}\right\} \quad (1)$$

Let us analyze the expression in parentheses: $\left(\frac{m\cdot g\cdot l}{H}+\frac{m\cdot(S_0\cdot\omega)^2}{2\cdot H}\right)=Q_p$.

The first term $\frac{m\cdot g\cdot 1}{H}=\omega_0$ is frequence of gyroscopic wobbling of gyropendulum, appearing because of torque due to gravity $m\cdot g\cdot 1$ and kinetic momentum of gyroscope H, the seconf term $\frac{m\cdot(S_0\cdot\omega)^2}{2\cdot H}=\omega_{0v}$ is angular rate of gyroscopic wobbling appearing because of vibrostabilizing moment $\frac{m\cdot(S_0\cdot\omega)^2}{2}$ and kinetic momentum of gyroscope H. Hence

$$Q_p=\omega_0+\omega_{0v} \quad (2)$$

The sum of these two terms (2) is frequence of gyroscopic wobbling of gyropendulum with down displaced center of mass, set on vibration base.
In compact form of general solution will be writtern as

$$\left.\begin{aligned}\vartheta=D\cdot sin\left[\left(\omega_0+\omega_{0v}\right)\cdot t+\varepsilon\right]\\ \psi=D\cdot cos\left[\left(\omega_0+\omega_{0v}\right)\cdot t+\varepsilon\right]\end{aligned}\right\} \quad (3)$$

From (3) gyropendelum is undergoing gyroscopic wobbling with amplitude D and frequency $Q_p=\omega_0+\omega_{0v}$. We can made a conclusion that the frequence of gyroscopic wobbling Q_p has increased in comparison with gyroscope on immovable foundation.

2.2 The experimental part

Complex, functional block diagram shown in Figure 1 was used to study gyropendulum with a upwards and down displaced center of mass.

Prototype of gyropendulum 1 was attached to the electrodynamic shaker type table 2. The table sets harmonic vibration at a fixed frequency. The amplitude and frequency of vibration are given by a signal generator 6.

Control over the vibration speed of the shaking table was made by piezoelectric sensors 3, their signals through a matching lock-in amplifier 4 [8-9] were supplied on the program-setting device 5. The last one are monitored and recorded the vibration parameters (amplitude and frequency).

The period of the precessional motion was measured manually using a stopwatch 7. The level of vibration produced in accordance with the previously performed calculations, based on parameters of real gyropendulum.

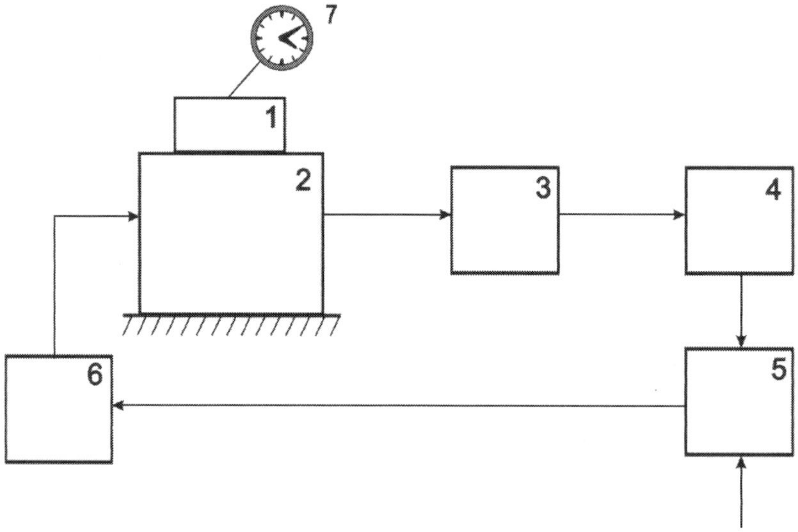

Fig.1. Functional block diagram.

3 Results and discussion

The main objective of the experimental studies was qualitative and quantitative evidence of change of speed of gyropendulum precession depending on the vibration parameters of the base (S_0; ω).

3.1 Gyropendulum with an upwards displaced centre of mass

This is a mode where gyropendulum with an upwards displaced center of mass located above the pivot point, on a vibrating base perfemes a processional oscillation with the same frequency as gyropendulum with a below the point of suspension to a fixed base was determined with the following vibration parameters: vibration frequency f = 130 Hz; vibration velocity V = 0.9 m/s; the total precession time of t=9 min; number of revolutions in time t of the experiment N = 3.

The period of gyropendulum precession of with a nupwards displaced mass center equals to 3 minutes. The mode in which the precession of gyropendulum absent, i.e. giromatic on a vibrating base acted like astatic gyroscope was fixed. This state of gyropendulum was recorded under the following vibration parameters: vibration amplitude S_0=1.2 mm; circular frequency ω = 580 rad/sec, i.e. speed of vibration is equal to $Vvib.$=0.7 m/s.

3.2 Gyropendulum with a down displaced centre of mass

Theoretical calculations have shown that the higher the dynamic force acting on gyropendulum is, other words the higher the frequency of vibration of the base (provided that the vibration amplitude constant), the higher the frequency of precession of gyropendulum is [2]. The experiments have confirmed the theoretical calculations. It was founded that the higher the ratio of dynamic forces and gravity (vibrational force) is , the smaller the period of the precession of gyropendulum is.

The recived results can be used in desing compensation scheme of action of vertical harmonic vibration on the gyropendulum.

4 Conclusions

It was found that the gyroscope with the displaced center of mass, located on a vibrating base, the precession rate depends on the amplitude and frequency (S_0; ω) base vibration (i.e, from the dynamic force) and also depends on the center of mass of the gyroscope location (above or below the point of suspension of the gyroscope). Such gyroscope changes its direction depending on the speed value (S_0; ω) vibration (i.e. dynamic force compared with the force of gravity). A case where there is no precession (dynamic force balances the force of gravity) was found.

Quantitative changes in the nature of the processional motion of gyropendulum depending on the vibration parameters were confirmed experimentally.

References

1. K.V. Lineitseva, V.I. Kopytov, *The 8th International Scientific and Practical Conference of Students, Post-Graduates and Young Scientists: Modern Technique and Technologies, MTT*, 59 (2002)
2. V.S. Volkova, V.I. Kopytov, *5th Korea-Russia International Symposium on Science and Technology*, 29 (2001)
3. V.S. Ivanova, 8*th International Scientific and Practical Conference of Students, Postgraduates and Young Scientists; "Modem Techniques and Technologies"*, 42 (2002)
4. E.V. Golodnykh, V.N. Borikov, *SIBCON*, 6693635 (2013)
5. A.V. Taracenko, V.S. Dmitriev, T.G. Kostuchenko, *KORUS*, 83 (2004)
6. A.N. Gormakov, *KORUS*, 20 (2004)
7. V.J. Skorykh, A.N. Gormakov, I.A. Plotnikov, *KORUS*, 42 (2001)
8. P.F. Baranov, E.I. Tsimbalist, V.N. Borikov, D.G. Soltanova, *IMEKO*, 116100 (2015)
9. P.F. Baranov, E.I. Tsimbalist, V.E. Baranova, *SIBCON*, 7146992 (2015)

Capabilities evaluation of spaceborne scientific equipment for geophysical applications

Ivan Kaloshin[1,*], *Vladimir* Kuznetsov[1], *Vladimir* Skripachev[1], and *Irina* Surovceva[1]

[1]Moscow Technological University (MIREA), 119454 Moscow, Russia

Abstract. The physical phenomena such as earthquake precursors recorded by on-board scientific equipment are examined in details. The selection method for geophysical satellite data of informative sources is discussed. The on-board scientific instruments complex application possibility for seismic activity diagnosis is shown. The development prospects of earthquakes precursors space monitoring system are defined.

1 Introduction

Satellites on-board scientific equipment provide information about various processes and phenomena on the Earth. Earthquakes are one of the most dangerous phenomena, due to the suddenness of their occurrence.

The importance of earthquakes precursors monitoring with on-board scientific equipment is determined by strict requirements for data acquisition operability, the earth coverage magnitude and data set sufficiency on their completeness (informativeness).

Strong crustal earthquakes formation process is accompanied by various precursors appearance: the Earth's crust deformation, electromagnetic field variations, anomalous cloud formations, etc. The main objective to identify earthquakes precursors from the space is the diagnosis of space and time environment parameters in seismically active regions.

We shall consider physical phenomena associated with seismic activity and recorded by on-board scientific instruments.

2 Physical phenomena and on-board instruments

The noise increase in the frequency range 100-500 Hz was recorded 14 hours before the earthquake from low orbit satellites over the epicenter of the earthquake with M>5.4. The abnormal increase of the low-frequency radio waves (0.1-15 kHz) intensity at the moment of earthquakes was recorded by magnetic and electric antennas [1].

According to [2] a correlation of the seismic activity with captured by the geomagnetic field particle fluxes variations was found. Spatial-temporal parameters analysis showed the rapid increase of the particles flux before the earthquake. This feature was formed during the phenomena distribution 2.5 hours before the earthquake strike.

* Corresponding author: kaloshin@mirea.ru

© The Authors, published by EDP Sciences. This is an open access article distributed under the terms of the Creative Commons Attribution License 4.0 (http://creativecommons.org/licenses/by/4.0/).

It is known that the earthquakes are accompanied by large thermal anomalies development that occur 2-3 weeks before the event and last 1-2 days after it. Such anomalies intensity is typically 3-4° C, with not only the earth's surface heating but the air heating above the epicenter as well [3].

Analysis of the cloud cover on the satellite images have demonstrated the presence of anomalous zones in the visible and infrared regions of the electromagnetic spectrum. The further studies have shown that the anomalous zone is best revealed in the fragments of space images of the earth surface areas sizes from 200 x 200 to 300 x 300 km [3].

There are basic requirements for all on-board scientific instruments: maximum efficiency; dimensions, weight characteristics and electromagnetic compatibility; to satisfy other requirements for mechanical, thermal, electrical and other satellites hardware characteristics. Now we shall briefly consider some scientific instruments of satellites for geophysical applications.

Ionosond is one of the effective means of scientific equipment to solve the problem of seismic activity precursor's detection. It spends pulse sounding of the ionosphere by providing timely information about its condition, including the critical frequency F2 layer ionospheric electron concentration. Use experience of such devices (e.g. IS-338) shows the possibility of operating on more than 300 fixed frequencies in the range of 0.3 ... 15.95 MHz for altitudes from 400 km to 1,000 km.

The detector of protons and electrons with the energy range of 20 keV - 2 MeV is widely used in the low orbit satellites. Such detector for measuring of the differential spectra of both electronic and ionic (proton) components of geoactive corpuscular radiation has been successfully tested on board the «Meteor» satellites.

Now some words about the global navigation satellite systems GLONASS/GPS signals receivers. These devices are essential for ionosphere radio tomography from low orbit satellites. It is also useful for radio occultation measurements. Those receivers are able to detect ionospheric disturbances.

To register anomalies in the ULF/VLF range it is necessary to install the instruments that allow to detect the signals appearing in the ionosphere before the earthquake. The instrument must work in the range of 10 Hz to 20 kHz, as the majority of the available experimental data has been obtained in this frequency range.

Functionally the radiometers consist of multiple spectral channels, both visible and infrared rangeы and provide images of the Earth surface. The processing images allow to obtain a number of precursors such as temperature anomalies, cloud cover anomalies and other meteorological precursors. As an example, it could be the radiometer AVHRR and radiometer MSU. In addition, these phenomena could be detected by passive microwave radiometers MTVZA with spatial resolution 20-100 km.

Mass spectrometers, which operation is based on ionized particles separation by their masses under the influence of magnetic and electric fields that allow the measurement of ionized and neutral components of ionospheric plasma. These instruments must provide the measurement of relative content of ion concentrations key component of the ionospheric plasma in the range of 1-65 mass amu with a resolution of 10-15 amu.

To estimate opportunities onboard scientific equipment of perspective satellites for short-term forecasting of earthquakes, it is necessary to choose the most informative sources of satellite data. To do that we have combined the expert information and paired comparisons methods. Table 1 shows the main sources of predictive information. The proposed method is based on the expert information processing.

3 Expert information processing technique

Each expert has filled in two matrices on the earthquake precursor's relevance and the spaceborne instruments feasibility. The count of experts N=6. The experts were employees of IKI RAS, MIREA, «Russian Space Systems» JSC, IRE RAS.

Weights calculation in accordance with the values, which ranged the matrices elements, is an iterative process [4]:

$$p^t = \mathrm{A}p^{t\text{-}1} \tag{1}$$

where p^0=(1,1,...,1)'.

During the iteration it is possible to obtain large weight values. To avoid that, the components of the vector p^t must be normalized at each step of iteration by following value:

$$\lambda^t = \Sigma_i p_i^t = \Sigma_i \Sigma_j a_{ij} p_j^{t\text{-}1} \tag{2}$$

According to the normalized factor the weights calculation procedure is:

$$P^t = \mathrm{A}p^{t\text{-}1}/\lambda^t \tag{3}$$

Table 1. Main information sources of earthquake precursor.

Parameter	Earthquake precursor information
A_1	Surface temperature
A_2	Temperature and humidity vertical profile
A_3	Cloud cover anomalies
A_4	Vapor flow
A_5	Upper-level wind (9-12 km)
A_6	Infrared radiation flows at the height of the upper clouds edge
A_7	Thermal anomalies
A_8	Local disturbances of ionospheric plasma (electrons concentration, electrons and ions temperature), including electron concentration vertical profile
A_9	Ion composition
A_{10}	ULF/VLF-emissions
A_{11}	Electromagnetic fields
A_{12}	Particles fluxes with energy spectrums

The computational process continues until the moment when the weighs obtained in successive iterations slightly differ from each other, i.e.

$$\max|p_i^t - p_i^{t\text{-}1}| < \varepsilon \tag{4}$$

where ε=0.005-calculation accuracy.

The priorities vector components estimation has been made separately. The polling of experts involved in the examination was carried out in view of the fact that their competence was the same. This procedure was repeated for each expert data processing. The final group score was obtained by averaging the individual assessments values.

Ranking was used to evaluate the significance of the experts polling results. Each value $A_1,...,A_{12}$ was assigned a rank - serial number in the ascending order.

To determine the strength of relationship between the ranks of the parameters $A_1,...,A_{12}$ a coefficient of concordance was determined. Since the ranks are disconnected, then the coefficient of concordance is defined as:

$$W=12S/[m^2(n^3-n)] \quad (5)$$

where m=2 – number of properties (relevance of the earthquake precursors, feasibility of the spaceborne instruments); n=12 – number of matrices parameters $A_1,...,A_{12}$;

$$S=\Sigma_m{}^n(\Sigma_j{}^m R_{ij})^2-(\Sigma_{i=1}{}^n\Sigma_{j=1}{}^m R_{ij})^2/n \quad (6)$$

where R_{ij} – rank of i parameter of j property.

The coefficient of concordance significance was estimated by means of Chi-squared statistic. Estimated value χ_c^2 in the absence of connected ranks is determined by the formula:

$$\chi_c^2=12S/[mn(n+1)] \quad (7)$$

As the defined value χ_c^2=6.90 exceed the table value χ_{tab}^2=4.57, with given significant level α=0.95 and degrees of freedom df=11, such coefficient of concordance becomes significant. Having applied the relative normalizing by maximum value, the following vector of priorities was obtained; the results are shown in Table 2.

To identify the priority informative sources of satellite data, the threshold with weighted average value L=0.7 was used (Figure 1).

Table 2. Normalized priorities vector.

A_1	A_2	A_3	A_4	A_5	A_6	A_7	A_8	A_9	A_{10}	A_{11}	A_{12}
0.684	0.658	0.776	0.387	0.338	0.365	0.962	1.000	0.779	0.889	0.630	).748

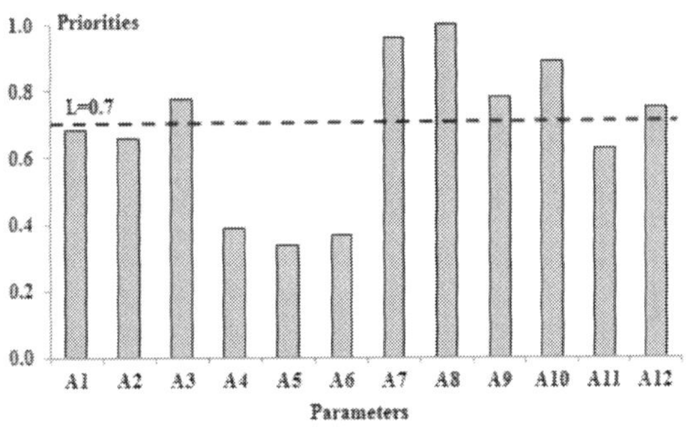

Fig. 1. Histogram of normalized priorities vector.

4 Results

One can see that parameters A_3, A_7, A_8, A_9, A_{10}, A_{12} have the highest priority for geophysical monitoring. According to the expert estimation results in order to detect earthquake precursors it is advisable to diagnose identified parameters and to have on-board scientific equipment for the registration following parameters: local disturbances of ionospheric plasma (electrons concentration, electrons and ions temperature), including electron concentration vertical profile; thermal anomalies; ULF/VLF-emissions; ion composition; cloud cover anomalies; particles fluxes energy spectrums.

Thus, the selecting method of the most informative sources of satellite data received by on-board devices should be based on the satellite dimensions, orbital constellation and above requirements. The main tools are:

- signals receiver of global navigation satellite systems (dual-frequency with a measurement rate of not lower than 20 Hz) for transionospheric sensing;
- ULF/VLF complex with 0.02-20 kHz work range;
- IR camera with spatial resolution of 15…250 m and resolution 5K;
- high-energy particle detector with a measuring range from 3 to 100 MeV for fluxes of energetic particles generated as a result of seismic activity.

5 Conclusion

The results obtained allow to estimate the optimum composition and technical characteristics of satellite on-board scientific equipment for sensing earthquake precursors, and to determine the prospects of the space seismic activity monitoring system development.

The following can be recommended as an important addition to satellite data for a forecast earthquake monitoring system: ground-based signals receivers of global navigation satellite systems (GLONASS, GPS) and ground-based VLF-receivers, which allow to observe seismo-ionospheric disturbances.

References

1. M. Parrot., ASR **15(11)**, 27 (1995)
2. V. Pustovetov, E. Ginzburg, A. Malyshev, I. Proshkina, Cosmic Research **31(3)**, 123 (1993)
3. A. Tronin, M. Hayakawa, O. Molchanov, J. Geodynamics **33**, 519 (2002)
4. V. Davnis, V. Tinyakova, *Prognoznye modeli ekspertnykh predpochteniy* (VSU, Voronezh, 2005) (in Russian)

Experimental investigation of frequency characteristics of the multistage slot membranes waveguide filters

Natalia Kopylova[1,*], *Alexei* Kopylov[1], and *Yuri* Salomatov[1]

[1]Siberian Federal University, 660079 Krasnoyarsk, Russia

Abstract. In this article, we present results of experimental investigation of amplitude-frequency characteristics (AFC) of the five-stage "filters with the waveguide-slot membranes" (FWSM) in the range of frequencies 5.4…8.5 GHz for six filters. Article shows the following trend change filter frequency response by increasing the width of the slots of the second and fourth membranes: decrease the quality factor, resonant frequency increase, reducing losses in the passband.

1 Introduction

There are a lot of papers of waveguide filters with metal membranes, which are arranged perpendicular to the direction of propagation of electromagnetic waves [1-5]. These works cover a large historical period: from 1946 [1] to 2000-ies [2-5]. These filters are called «iris filters» ([3-5]). However, it is more convenient to name them "filters with the waveguide-slot membranes" (FWSM). Using filters of this type allows getting wide bandwidth or attenuation with low weight and sizing characteristics of the filter. Taking such filters is the most important in the centimeter wavelength range of the microwave to minimize their size in comparison with conventional filters on long line segments.

The results of experimental studies of amplitude-frequency characteristics (AFC) of one of these types of filters in the frequency range 5.4 ... 8.5 GHz we present in this article.

2 Filters design

The structure of the investigated five-stage FWSM is schematically shown in the Figure 1. Figure 1a) shows generalized filters topology, Figure 1b) – filters on the side.

Our five-stage FWSM contains a set of five waveguide-slot membranes with "B" wide of slot (Figure 1a)). The value "B" is constant for the first, the third and the fifth membranes. For the first and for the fifth membranes, it is equal to 22 mm and for the third membrane– 15 mm. For the second and for the fourth membranes value "B" is 1 mm, or 2 mm, or 3 mm, or 4 mm, or 5 mm, or 6 mm. Slot is arranged parallel to broad wall of the waveguide. The cross section of the waveguide is 35x15 mm. The thicknesses of the filter

* Corresponding author: kopaph@yandex.ru

© The Authors, published by EDP Sciences. This is an open access article distributed under the terms of the Creative Commons Attribution License 4.0 (http://creativecommons.org/licenses/by/4.0/).

 DOI: 10.1051/matecconf/201710201025

membranes are: 6 mm for S1 and S5; 10 mm for S3; 0.25 mm for S2 and S4 respectively (Figure 1b)).

Thus, the design of the FWSM is a symmetrical structure consisting of five cascade waveguide-slot membranes with different membrane thicknesses and slots width. The lengths of the slots of the membranes are chosen equal to the length of the broad wall of the waveguide. All slot membranes disposed along to the common (based) side of the waveguide (Figure 1b) with a bold type cross-hatching line).

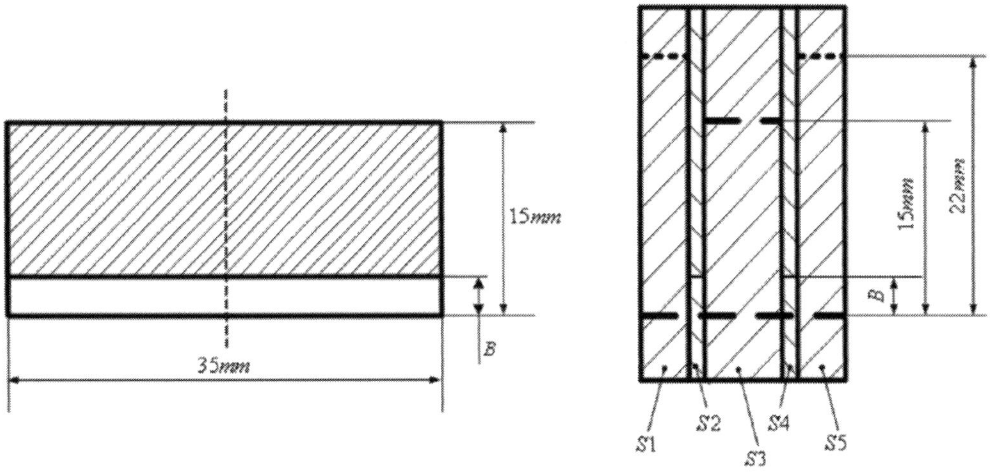

Fig. 1. The structure of the investigated five-stage "filter with waveguide-slotted membrane" (FWSM): a) Generalized filters topology, b) Filters on the side.

The filter design was selected by us experimentally to determine the form of the amplitude-frequency characteristics of the filter in bandwidth and influence of the width of slots of waveguide-slot membranes.

3 Results and discussions

For our investigations, we made six FWSM with structure shown in Figure 1 and different wide of "B" slot. The value of "B" was for the first filter equal to 1 mm, for the second filter – 2 mm, for the third filter – 3 mm, for the fourth filter – 4 mm, for the fifth filter – 5 mm, and for the sixth filter – 6 mm.

Figure 2 demonstrates amplitude-frequency characteristics (AFC) of all six filters investigated by us. Figure 2a) shows the AFC of the first FWSM which value of "B" is 1 mm; Figure 2b) – the second FWSM which value of "B" is 2 mm; Figure 2c) – the third FWSM which value of "B" is 3 mm; Figure 2d) – the fourth FWSM which value of "B" is 4 mm; Figure 2e) – the fifth FWSM which value of "B" is 5 mm; and Figure 2f) – the sixth FWSM which value of "B" is 6 mm.

Course of frequency dependencies of transmission filters' coefficient in Figure 2 proves: investigated filters allow to implement filters of AFC type low-pass filter (LPF) (Figure 2a)); filters with AFC type of band-pass (BPF) filter in the Figure 2b), 2c), 2d), 2e) and AFC-type high-pass filter (HPF) in the Figure 2f). Moreover, our FWSM's exhibit the properties of band-stop filter. This is observed at frequencies above 7 GHz in filters, which frequency responses are shown in Figure 2a), 2b), 2c), and in filter with frequency response is shown in Figure 2f) at frequencies less than 6 GHz.

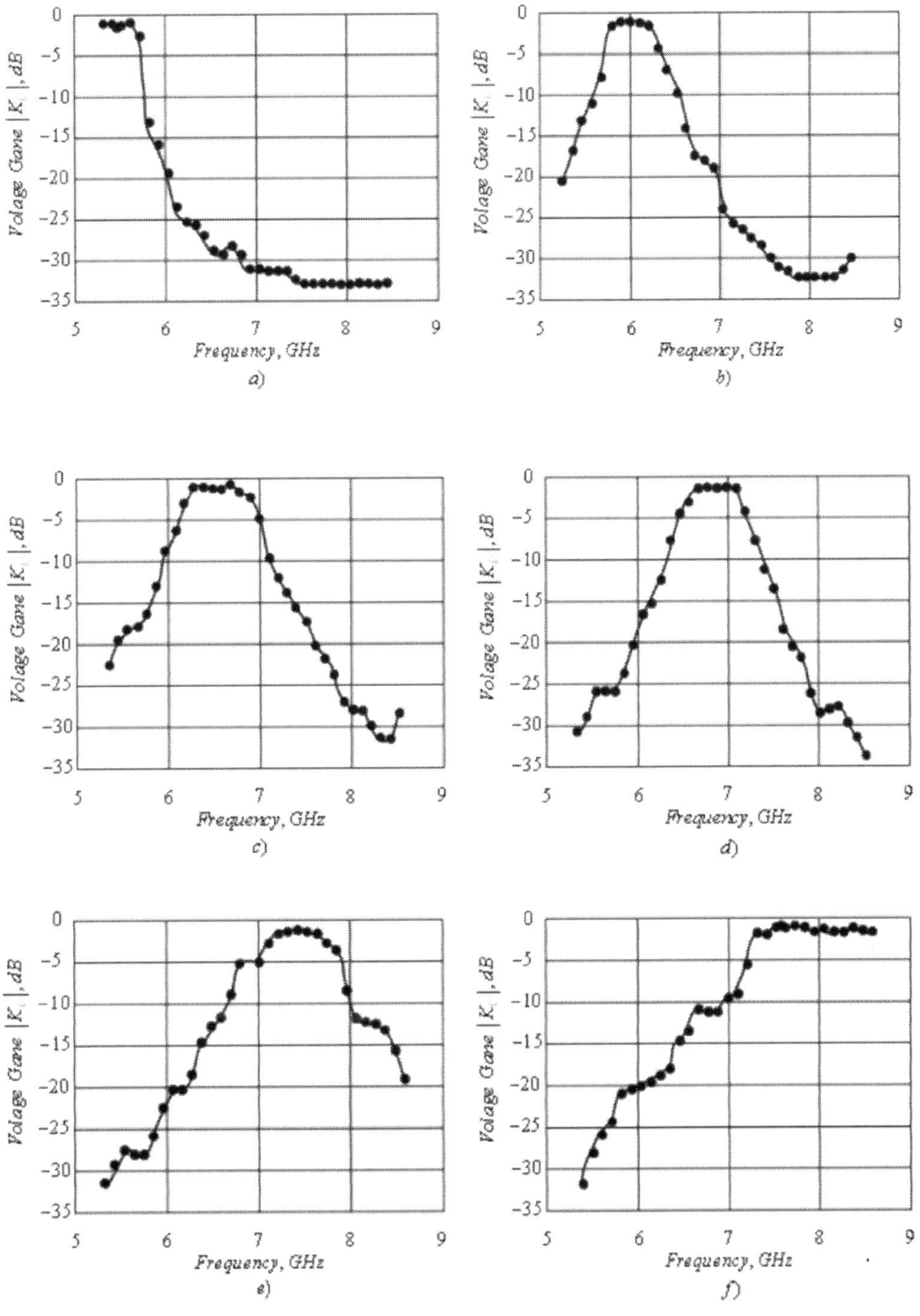

Fig. 2. Amplitude-frequency characteristics of all six filters investigated by us: 2a) AFC of the first FWSM; 2b) AFC of the second FWSM; 2c) AFC of the third FWSM; 2d) AFC of the fourth FWSM; 2e) AFC of the fifth FWSM; 2f) AFC of the sixth FWSM.

Besides, the band-pass has been expanding with increasing "B" magnitude of gap in the second and fourth filter membranes. At the same time we notice decreasing slope of the filters' frequency characteristics. Both of going facts are easily explained by increasing of bond between filter elements with increasing the "B" value of second and fourth membranes' gaps. We notice the values of insertion loss in the bandwidth for all investigated filter is up to 0.1 dB, which is the limit of the measurements' accuracy as well.

In addition, we should notice a significant change in the minimum attenuation of AFC filters on frequencies: changing of the "B" amount from 1 mm to 6 mm in the second and fourth filters membranes comes to changes of these frequencies' value from 5.5 GHz to 7.5 GHz.

4 Conclusions

We believe these results are useful for realization of waveguide filters with different frequency response types for some areas such as radio communication, radio navigation and radar equipment of wide applications.

References

1. R.M. Fano, A.W. Louson, RLE, Mass. Inst. Techn. **8**, 1 (1946)
2. J. Carbonell, J. Roglá Luis, V. E. Boria, D. Lippens, IEEE Trans. **4**, 1527 (2006)
3. I. Stevanovic´, P. Crespo-Valero, R. J. Mosig, IEEE Trans. **1**, 189 (2006)
4. Q. F. Zhang, Y. L. Lu, IET Microw. Ant. Prop. **4**, 1256 (2010)
5. C. A. Leal-Sevillano, J. R. Montejo-Garai, J., A. Ruiz-Cruz, J. M. Rebollar, IEEE Trans. **3**, 1 (2016)

The electric drive control system of corner reflectors of the spacecraft interferometer

Sergey Langraf[1,*], *Dmitriy* Bunkov[1], *Ivan* Odnokopylov[1], and *Olga* Galtseva[1]

[1]Tomsk Polytechnic University, 634050 Tomsk, Russia

Abstract. This paper presents the electric drive control system of corner reflectors of the infrared Fourier transform spectrometer for meteorological satellite. Limited-swing brushless DC motor with a torsional bearing is used as a drive motor. Stringent requirements are imposed to the drive control system for stabilization of moving speed of the reflectors while obtaining the interferogram and for limitation of the reverse time. Research of influence of torsion on steady-state error of speed and ways of its compensation is conducted. The obtained results are consistent with the simulation results of the drive. It is shown that the developed drive is operable in a spacecraft system.

1 Introduction

Electrical engineering is widely used in the space industry. There are several features associated with toughening of requirements for such electromechanical devices: creation of compensating device, special conditions to the heat sink, the high requirements for reliability and fault tolerance, radiation resistance.

The creation of meteorological satellites is one of the promising directions of development of space industry for civil purposes. They allow observing the Earth's atmosphere from space. Due to the rapid development of space meteorology in 2014 the infrared Fourier transform spectrometer (IFTS) appeared on board of the spacecraft "Meteor-M". Thanks to the advent of IFTS it has become possible to obtain an infrared spectrogram of the atmosphere and the Earth's surface that facilitates the temperature and humidity sounding [1, 2].

The scan accuracy of infrared radiation depends on the accuracy of the reciprocating motion of drive of the pendulum of corner reflectors of the interferometer; this is the only constantly moving element of IFTS. Therefore, an important task to develop the drive control system that meets the high requirements to IFTS for obtaining the interferograms in space conditions [3].

*Corresponding author: langraf@tpu.ru

© The Authors, published by EDP Sciences. This is an open access article distributed under the terms of the Creative Commons Attribution License 4.0 (http://creativecommons.org/licenses/by/4.0/).

2 Materials and methods

The operation principle of IFTS is based on the phenomenon of interference of waves. The interferogram from the photodetector is a function of the optical difference of path of two beams [4]. The division into two equal intensity beam occurs using the beam splitter (Figure 1). The change of optical difference of path transpires due to the uniform displacement of the pendulum with corner reflectors on the value of ±2 degrees. Stringent requirements are imposed to the drive of the pendulum of reflectors. Firstly, it is necessary to ensure a high stability of defined speed of at least 1% of the operating speed at moving of corner reflectors. The quality of the resulting interferogram depends on the accuracy of sustaining of the defined speed directly. Secondly, it is essential to make a quick change of direction movement of corner reflectors at 0.05 s. This will reduce an idle of IFTS operation.

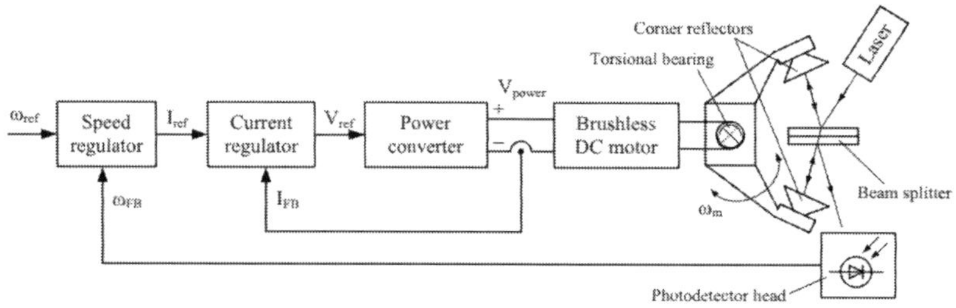

Fig. 1. Functional scheme of the electric drive of corner reflectors of the interferometer.

The electric drive system works in a cyclical pendulum mode, consisting of two zones: interference and reverse. Processing the infrared radiation from the Earth's surface occurs in the interference zone. The drive should work in mode of speed stabilization at the approach to the interference zone after each reverse.

Limited-swing brushless DC motor is proposed to use for implementation of the electric drive system [5]. Ensuring the long-term operation of the limited-swing drive in space conditions is achieved using the torsional bearings which have significant advantages compared to radial bearings.

For implementation of the requirements of the speed stabilization it is need to use a double circuit control system made in accordance with the principles of the subordinate regulation. Formation of the feedback signal for the speed loop occurs by means of high-precision laser measurement system of moving of corner reflectors. An additional laser interferometer is used for this purpose; laser radiation passes through the beam splitter and corner reflectors and gets to the receiving photodetector (Figure 1). The resulting sequence of pulses converted into signals with accurate information about the current speed of movement on the photodetector output.

The control loop by motor current is the inner loop of control system, the control loop by speed of the moving of the drive is the external circuit of system. Proportional-integral controller is used in current loop. The optimization is performed in view of achieving the limit rapidity of the system; it is characterized by a total inertia of the elements of the power channel and current measurements. Also proportional-integral controller is used in the control loop by speed of the moving for improving the accuracy characteristics. The optimization is performed according to the method of symmetrical optimum [6] to ensure a maximum rapidity with minimizing the error value.

3 Results

Analysis of conditions of drive operation with the torsional bearings shows that a gradual increase of resistance torque from the torsional bearings occurs at the deviation of the pendulum with the reflectors from the concerted position. Thus, disturbing load from the torsions is changed by the linearly-increasing dependence at the drive movement with constant speed [7].

Mathematical analysis of transfer functions of electric drive control system shows that the use of proportional-integral controller in speed loop gives the opportunity to provide the second-order astaticism on control for electric drive; only the first-order astaticism on disturbance is possible. In this regard, the steady-state error of regulation will appear during the electric drive operation with a constant speed of movement in conditions of linearly increasing resistance [8]. The calculated value of the steady-state error, as a result of exposure of linearly increasing load from the torsions, can be determined from the expression of the transfer function of loop speed on disturbance.

The calculated value of the steady-state error on speed during obtaining of the interferograms was defined on level $\Delta_{calc}= 9.54 \cdot 10^{-3}\%$, which is not beyond the initial requirements to the drive of movements of corner reflectors.

4 Discussion

Functional testing the proposed control system was performed using a drive simulation of corner reflectors (Figure 2).

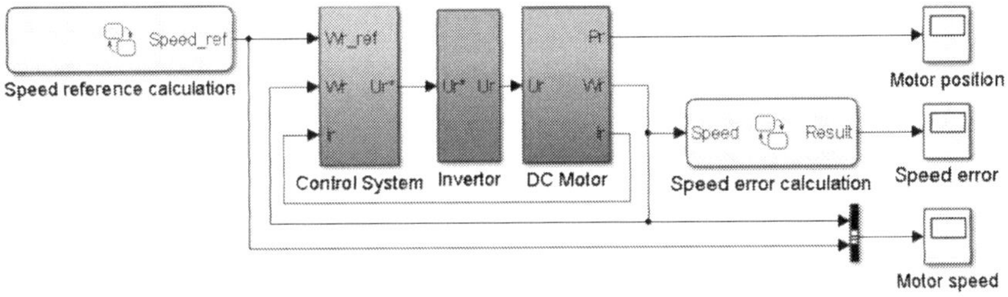

Fig. 2. Simulation model of electric drive of movement of corner reflectors of the interferometer.

The model was developed in software Matlab Simulink and takes into account the properties of the autonomous voltage inverter with limited-swing brushless DC motor with excitation from permanent magnets [9]. Power-up sensor with the S-shaped profile is used for the formation of smooth movement and optimal reverse of the motor, for limitation of the maximum speed up and for exclusion of jerks at the electric drive operation. The main feature of this model is the consideration of the torque load from the torsional bearings depending on the current angle of the pendulum of the interferometer.

The main simulation results are presented as graphs of transient processes (Figure 3).

Drive speed error has constant level at work in the interference zone. The error sign depends on the current direction of movement, but module of error value corresponds to the expected calculated value at the level $\Delta = 9.55 \cdot 10^{-3}\%$. In the reverse zone, the electric drive shows high dynamics; the time of readiness for the next cycle of work in the interference zone is $t = 0.05$ s.

It is shown that the developed system of the pendulum electric drive with corner reflectors can be successfully used for organization of the IFTS work in the composition of the spacecraft.

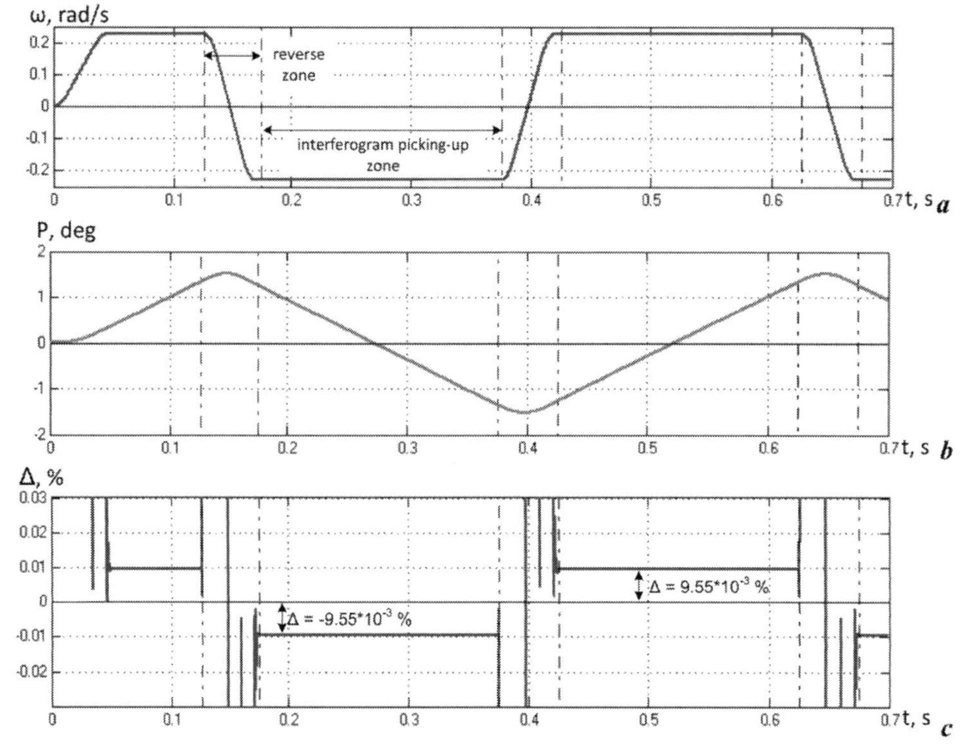

Fig. 3. Graphs of transient processes of electric drive of corner reflectors, where *a* is a speed, *b* is the position, *c* is the speed error.

5 Conclusions

The results of the study show that a module of speed error of drive has a constant level and corresponds to the expected calculated value at the level Δ= 9.55·10-3 % compared to the working speed. It corresponds to the requirements for IFTS of the electric drive in the obtaining mode of interferograms (1%). In the reverse mode the drive is ready to the next cycle for work in the interference zone at t=0.05 s; it fully meets the necessary initial conditions.

References

1. A.G. Efimov, V.F. Panin, V.O. Los, *Proc. of CriMiCo*, 8087502 (2003)
2. H.M. Gong et al, *Proc. of IRMMW-THz*, 9693879 (2006)
3. R.L. Ward et al, Class. Quantum Grav. **31(9)**, 095015 (2014)
4. A.V. Velasco, *Proc. of PIERS*, 16481423 (2016)
5. A. Jagadeeshwaran et al, Aerospace and Electronic Systems Magazine **31(6)**, 16105694 (2016)
6. S.V. Langraf, A.Y. Alyokhin, *Proc. of MTT*, 7341914 (2001)
7. S. Xunwen, W. Shaoping, Zh. Dongmei, Sh. Jian, *Proc. of PHM*, 11140859 (2010)
8. O.M. Lysova, T.G. Nesterenko, I.V. Plotnikova, *Proc. of ICINS*, 23 (2008)
9. P. Murali Krishna, N. Kannan, Proc. of PEDES, 5391142 (1996)

Design of a kilowatt DC-DC converter

Hongxing Liu[1], *Chenwei* Fu[1], *Na* Li[1], *Soldatov* Alexey[2], and *Wei* Han[1,*]

[1]Jilin University, College of Physics, 1505 Linyuan Road, Changchun, China
[2]Tomsk Polytechnic University, 634050 Tomsk, Russia

Abstract. In view of the low power of traditional DC-DC converters, a DC-DC converter with a kilowatt power is designed. The input signal's frequency is 1 kHz and the duty cycle is 5%. The PWM signal controls the high-speed conduction or cut-off of the switch tube. The input DC voltage is 36V, and the output voltage is twice as high as the input voltage. The output power is greater than 1 KW; the circuit conversion efficiency is 87.21%.

1 Introduction

At present, DC-DC converter mainly works on a low-power voltage doubling, voltage inversion and other aspects. The mainstream applications are the battery-powered mobile phones, digital cameras and other fields to drive LED backlight which should be provided with appropriate power [1-3]. Another very extensive application is EEPROM and flash memory to provide read and write power, the memory of the power rail is generally 1.8 V, 3.3 V, as read needs +5 V and erase needs +12 V, that's why we integrate the DC-DC converter into the IC for single-supply operation [4-7].

For the current charge pump applications, the main problem is the difficulty to achieve high-power, high-voltage applications, as high-power aspects of inductive switching DC-DC converter occupies an absolute advantage [8-11]. But the charge pump has the advantages of small volume, high efficiency, low quiescent current and low minimum operating voltage, so it is more important to design and manufacture a high-power charge pump [12, 13]. In this paper, the circuit is designed and produced by MATLAB simulation software. The output power of the charge pump is more than 1 KW and the conversion efficiency is 87.21% by PWM signal controlling the switch while charging and discharging the capacitor.

2 DC-DC converter design

2.1 DC-DC converter twice the boost principle

The principle of the DC-DC converter double boost shown in Figure 1(a), using capacitor charge and discharge to achieve an output voltage that is twice the input voltage. First,

[*]Corresponding author: whan@jlu.edu.cn

© The Authors, published by EDP Sciences. This is an open access article distributed under the terms of the Creative Commons Attribution License 4.0 (http://creativecommons.org/licenses/by/4.0/).

close the *S1*, *S2*, and *S3* switches, and the capacitors *C1* and *C2* (called *C1* and *C2* for the pump capacitance) are charged in parallel. The DC power supply charges the capacitor almost instantaneously. Then, *S1*, *S2* and *S3* are turned on while the other set of switches *S4* and *S5* are closed, and *C1* and *C2* are in series to provide power for *C3* (called the load capacitance).

Under ideal conditions, the two pump capacitors are connected in parallel to a power supply in order to charge until voltage is 36 V. The next stage is to connect the capacitors in series in order to charge it till 72 V, in order to achieve a double boost through doubling the input voltage.

2.2 Design of MATLAB simulation circuit

As shown in the following circuit diagram, MATLAB software has been used to design and simulate the circuit. The power supply is a DC power supply and is set to + 36 V. Switch can be used as a general-purpose switch, respectively, for each switch input phase reversed PWM control signal. A PWM control signal having a duty ratio of 5% is inputted for *S1*, *S2* and *S3*, and a PWM control signal having a duty ratio of 95% is inputted to *S4* and *S5*, with a delay of 0.000005 second from the previous signal, to realize controlling the two sets of switches at the same time. The control signal diagram is shown in Figure 1(b).

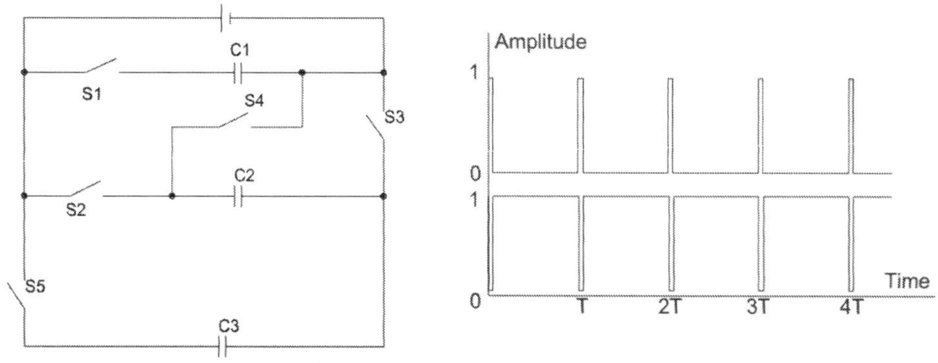

(a) (b)

Fig. 1. (a) DC-DC converter twice the boost principle, (b) Input PWM control signal.

These two processes complete one charge and discharge process, the whole process is completed in a PWM cycle. The capacitor charging formula is as follows:

$$V_t = V_u \cdot \left[1 - \exp\left(-\frac{t}{RC}\right)\right] \tag{1}$$

Where, V_t means the voltage across the capacitor for time t, V_u means the voltage after the capacitor is fully charged, *R* is the load resistance value and *C* is the charging capacitance value, which is calculated by the formulas:

$$t = 2RC, V_t = 0{,}86V_u \tag{2}$$

$$t = 3RC, V_t = 0{,}95V_u \tag{3}$$

$$t = 4RC, V_t = 0{,}98V_u \tag{4}$$

$$t = 5RC, V_t = 0{,}99V_u, \tag{5}$$

Visibly, after $5RC$ time of the charge, the capacitor is almost full charge. Discharge formula is as follows:

$$V_t = V_u \cdot \exp\left(-\frac{t}{RC}\right) \tag{6}$$

When the time t tends to infinity, the voltage is reduced to zero. The charge of the load capacitance is required to be charged more than the pump capacitance with respect to the charging process, so that a PWM control signal having a duty of 5% is inputted to *S1*, *S2* and *S3*, and a control inverting signal is inputted to *S4* and *S5*. As the capacitor charge and discharge processes are nonlinear, processes needs to be repeated in order to achieve the accumulation of charge, that is, to achieve the capacitor's charging.

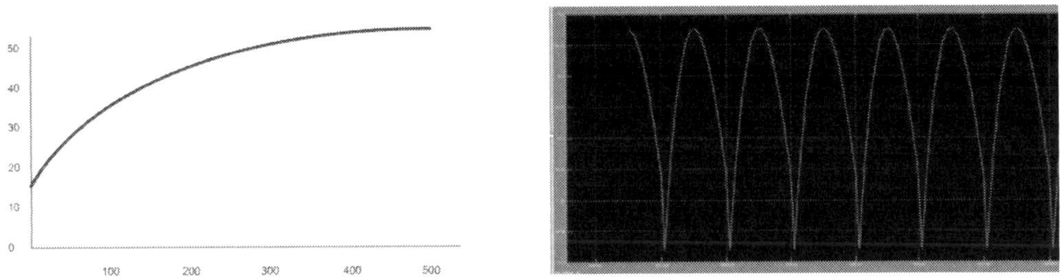

(a) (b)

Fig. 2. (a) Output voltage waveform before filtering, (b) Filtered ripple waveform.

The capacitance of each pump capacitor is 250 V, 2200 μF and 0.01 Ω, which is equivalent to 250 V, 550 μF and 0.0025 Ω. The internal resistance of shunt capacitor is reduced, which can greatly reduce the loss of the circuit. Maxwell load capacitors were selected at 2.7 V, 350 F, 0.1 Ω super capacitor, the two capacitors in series, making it increase in the voltage across the capacitor.

In order to test the output power of the load capacitance, load resistance is added to the load capacitance's circuit. Testing the value of voltage and the current on the branch of the load resistance circuit can provide the output power of the load capacitance.

After the simulation test, it has been observed that the output voltage ripple is large and in order to obtain an ideal DC voltage, a filter is added to the load capacitance of the output voltage. The filter inductance value is selected at 3.3 mH. A very large inductor value will increase the circuit resistance, and thus increase losses. The ripple factor after filtering is 0.35 V, the ripple coefficient is reduced to 0.62%, and the ripple factor satisfying the circuit design requirement is less than 1%. Before and after the effect of filtering comparison chart is shown in Figure 2(a) and Figure 2(b).

3 Results and analysis

Figure 3 shows the simulation results where the input frequency is1 kHz, duty cycle is 5% of the PWM signal. From the simulation results, it can be seen, that the final output voltage is 62.79 V and the output current value is 20.93 A. The output power is $P = UI$ = 62.79 · 20.93 = 1314.19 W, to achieve the purpose of designing the circuit. The circuit is a two-fold boost circuit, the ideal output voltage of 72 V, because of the presence of internal resistance, there is a certain loss, so conversion efficiency η = 62.79 / 72 · 100% = 87.21%, higher than the traditional charge pump conversion efficiency.

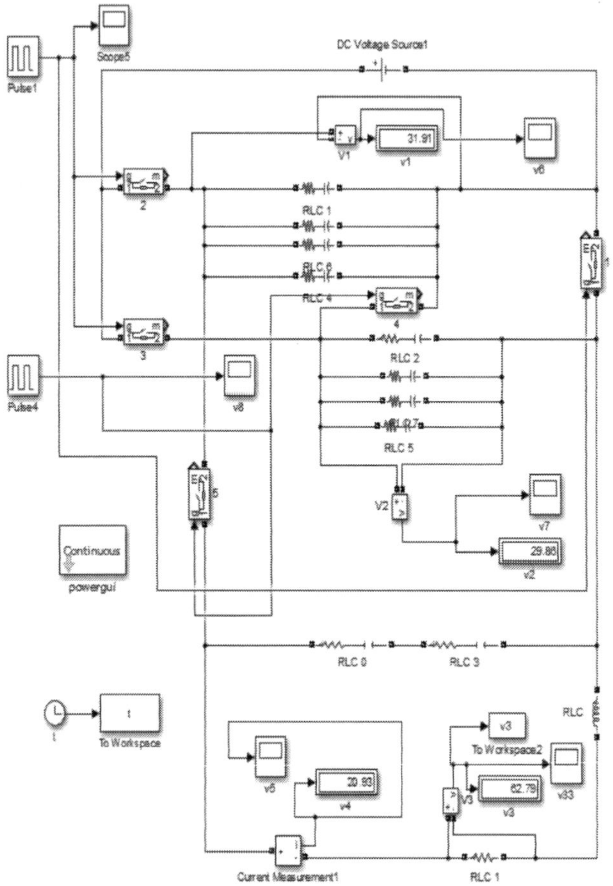

Fig. 3. MATLAB simulation results.

4 Conclusions

In order to solve the problems of current charge pump, the simulation of the software is carried out by MATLAB. The output voltage of the charge pump circuit is twice as large as the input voltage and the output power of the circuit is more than 1 KW by selecting the components and designing the circuit. This paper is based on the advantages of DC-DC converter, the design of lower cost, higher output power, higher conversion efficiency of the charge pump, breaking the traditional charge pump low power limitations.

References

1. Z Jinqing et al. Mod.Electr. Tech **37**, 148 (2014)
2. J Kai, Z Ying, Microelectr J **45**, 599 (2015)
3. F. Canales et al., IEEE Trans. Power Elect. **17**, 898 (2002)
4. E. Juarez-Hernandez, ICECS **1**, 349 (2001)
5. J.A. SABATE et al., IEEE APEC, 275 (1990)
6. M.J. Schutten, IEEE Trans. Power Elect. **18**, 659 (2003)

7. L. Mateu, F. Moll, SPIE **5837**, 359 (2005)
8. Chao Lu et al., DATE, 1602 (2010)
9. J. F. Dickson, IEEE J Solid-State Circuits **11**, 374 (1976)
10. Jieh-Tsorng Wu, Kuen-Long Chang, IEEE J Solid-StateCircuits **33**, 592 (1998)
11. Grigorev A.P. et al., *Mod Tech+Technol* **896064**, 97 (2000)
12. A.I. Soldatov et al., *SIBCON*, 7491869 (2016)
13. E.Y. Burkin, O.A. Kozhemyak, Instrum Exp Tech+ **59**, 245 (2016)

Does a University graduate need a portfolio?

Kseniya Mertins[1,*], *Veronica* Ivanova[1], *Yury* Daneykin[1], and *Natalia* Daneikina[1]

[1]National Research Tomsk Polytechnic University, 634050 Tomsk, Russia

Abstract. Every year an increasing number of employers get interested in the university graduates, most suitable for the requirements the companies set. However, the standard CVs often do not reveal the potential and peculiarities of formation of graduate competences. The authors suggest using portfolio instead of a CV. The authors analyzed 1797 portfolios of TPU students and offered optimal structure and criteria of portfolio. In addition, the authors provide an example of how the achievements reflected in portfolio can be considered. The conclusion is made about the necessity of the portfolio. The authors determined employer's role in the formation and evaluation of the portfolio. Moreover, the value of tracking the dynamics of growth of portfolio components is denoted as a new performance evaluation system in higher education.

1 Introduction

Modern world has been a global market for quite a long time already. Education is also a product sold around. Therefore, universities around the world should analyze the needs of the modern world to timely offer what students and industry demand [1]. Within two years, students have been enrolled in Russian universities with consideration of the results of their individual achievements, i.e. portfolio compiled during the last school year. The interest of enterprises in the formation of graduates' portfolio is also worth noting. Now the questionnaires offered by employers ask to describe the achievements of graduates in science and social activities apart from GPA - grade point average. The authors consider it possible to continue the practice of formation and evaluation of the portfolio documents in the period of study at the university. This on the one hand, will serve as an additional motivation to the active life position of the students, beginning with the first day at the university, and on the other hand, will help to ensure a more profitable positioning of a graduate (as it contains much more information for employers than the standard CV does).

Many researchers consider the portfolio as a test of formation of graduate competences during training period. According to some authors portfolio allows solving two main tasks: to track individual progress that a student has made in the process of education without direct comparison with the achievements of other students and to evaluate their educational achievements and supplement (replace) the results of testing and other traditional forms of control. In this case, the outcome document of the portfolio can be viewed as an analogue of a testimonial or of the certificate of test results (or serve along with them).

* Corresponding author: mertinskv@tpu.ru

© The Authors, published by EDP Sciences. This is an open access article distributed under the terms of the Creative Commons Attribution License 4.0 (http://creativecommons.org/licenses/by/4.0/).

2 Portfolio peculiarities

Portfolio of the university graduate is a set of documented learning outcomes (including self-development) submitted for external evaluation and demonstration.

Types of portfolios can be divided according to the type of student activities: study, research, public, etc.

Adriana Baez, Wendy Pacheco and Caroline B. Appleyard distinguish [1] three types of portfolios (career, assessment, and developmental) depending on a variety of functions.

Career or professional portfolios are used to organize, track, store, and maintain documents related to their training and career, included skills, experience, and samples of work or accomplishments, which will supplement the curriculum vitae (CV). Most often, this type of portfolio is made to advance the student who already has experience up the career ladder. Assessment portfolios are only a part of educational program, which provides objective assessment of the graduate learning outcomes for one or another discipline. In addition, this type of portfolio is promising in terms of transition from listing grades in diplomas of higher education to the disclosure of specific skills by giving proof to the level of their formation (the level of formation of competences).

Based on the literature review, the authors believe that portfolio should show different areas of graduate's activity from professional in a certain field to the achievements in the direction, which they choose for self-development. Types of student activity considered in portfolio (learning, research, management, social activities, etc.) are presented in the form of portfolio structure, thus forming a complete image of a graduate.

The portfolio should include nine portfolio parts shown in Table 1, showing individual features of the university graduate. Each part corresponds to results, assessment part and assessment object. Part 1, named «Personal information» is provided by curriculum vitae. A curriculum vita is typically sent with a job application. Educational part describes education (academic diplomas modules) and training, hard skills and soft skills (the growth of the student through the educational program). The authors consider it important to supplement academic diplomas modules with enumeration of leading lecturers, scientific adviser of qualification work, as well as consultants. This improves the social responsibility of the subjects of educational activities and is beneficial for University image. Part 2, named «Educational part» includes educational projects (Description and achievements) according to CDIO initiative [3-6].

In addition, the portfolio may be completed by the 10th part - self-evaluation (self-evaluation of soft and hard skills plays a certain role in the formation of the future expert), if such skills as reflection, self-analysis etc. are important for employers. Probably, it will be important for graduates is to tell about themselves - their mission, goals, etc. for different stakeholders of portfolio [7].

Stakeholders have different purposes for creating and maintaining of portfolio and the following are among them:

– for employers - the formation of personnel reserve, reduction of adaptation period on the enterprise;

– for a graduate - systematization of individual achievements, designing of individual educational and career path, building map of competences;

– for a student - choice of development path in the university, opportunity to assess the potential for participation in competitions and scholarship programs, self-evaluation and career planning;

– for the university administration - automation of the selection of candidates for a variety of grants, scholarships and competitions, systematization of data on the various student activities, the possibility of scheduling PHD personnel reserve;

– for departments - use of portfolio as a tool to rate students.

However, despite different objectives there is something in common that brings together all stakeholders and this is an objective assessment that allows completing the most systematic image of graduates having specific competences.

Table 1. Components of a portfolio.

N п/п	Portfolio part	Results	Assessment part
1.	Personal information	Curriculum Vitae	-
2.	Educational part	Education (academic diplomas modules), Certificates, Certificate of Higher Education, Web	GPA - grade point average Skills
		Training	Skills
		Projects (Description and achievements) [2]	Skills Experience
3.	Professional part	Experience, Cases, Internship	Skills
4.	Scientific part	Research Projects (Description, achievements and awards)	Skills
5.	Social part	Projects (Description and achievements)	Skills Results
6.	Management part	Personal mobility (universities transcripts), MOOCs, Open University certificates, Language	Skills Level of language knowledge
7.	Entrepreneurial part	Start-up, Patients	Skills, Results
8.	Professional mistakes	Examples of failures where solutions were found	Experience
9.	Publication activity	Publications (published articles)	Skills

3 Methodology of portfolio assessment

Portfolio, as a system of evaluation of educational achievements generally focuses on the assessment methodology. The authors emphasize the necessity to monitor the dynamics of the activity of university graduates using presented portfolio (dynamics of change of portfolio performance - to evaluate progress and assess mastery of skills and concepts or completion of specific tasks and assignments. Employers are involved in portfolio assessment being one of portfolio customers. In addition, employers can comment on the graduates, assess the level of formation of competences, as well as to evaluate the level of scientific work.

Who is involved in the formation of the portfolio? The stakeholders that take part in the formation of the portfolio. Participants may vary depending on the specialization of the university, student activities, logistics and ecological environment within the university, practice-oriented training of university graduates, etc.

Portfolio can be used for student's assessment, including competition of the scholarship programs of enterprises.

Section "Publication activity", which may be one of the criteria for assessment of the applicant for a scholarship from the company, can be assessed using TPU system

«Flamingo», which helps to rank publications by significance, by type of publication, etc. Presentations at international conferences weigh 12 points, presentations at Russian conference are given 10 points, conference abstracts - 5 points, articles in scientific journals indexed by Web of Science and Scopus international database - 75 points, articles in peer-reviewed journals from the list of HAC, indexed by Russian database "Russian Index of Scientific Citation" - 75 points. Thus, it is possible to select which parameters are suitable for selection in each scholarship competition.

In addition, the following criteria are proposed for portfolio:

1. Online access.
2. Relevance of sections to information requested by employers.
3. Consistency, completeness of the description of achievements.
4. Using the rules of grammar and punctuation, language (writing skills).
5. Usability.
6. Confirmation of the results by a tutor, scientific supervisor, other stakeholders.
7. The possibility of evaluation by external experts in the process of work with the portfolio (evaluation by accredited experts).
8. Integration with University databases (including feedback).
9. Registering the dynamics of indicators growth in the framework of portfolio components (time span - annually).

4 Results: statistics

Analysis of work in the system "Flamingo" in TPU confirms the necessity of portfolio management. In April 2016 1797 portfolios (without division into bachelors, masters and PhD students) were recorded.

Participation of students in 94 competitions of scientific-research work and scholarship programs, 12 Olympiads and competitions of educational achievements were organized from February 1, 2015 until March 1, 2016.

More than 7000 students of 1-4 years of study and more than 1,900 Master Degree students do research work.

More than 6000 students of TPU take part in practical training annually, more than 67 percent of students do practical work in organizations and enterprises.

More than 25 percent of the results performed in students' qualification works, including more than 36 percent of the results of Masters' qualification works are recommended for implementation due to the practice-oriented approach to the implementation of work.

TPU plans to issue portfolio as Diploma Supplement, modifying «Flamingo» system, adding existing sections on the scientific and educational activities.

Thus, TPU can help personnel services of enterprises in the study of documents of more than 1,500 graduates. More than 1,500 graduates will annually receive a diploma supplement in the form of portfolio, including the passport of formed competencies, achievements in scientific, educational, creative and sporting activities.

5 Conclusions

Portfolio allows considering not only the level of professional competences of students, but also the level of comprehensive self-realization of students in educational environment. The introduction of the portfolio in the university allows not only using of modern technologies in training and education, but also developing of motivation to competence, linguistic and professional mobility.

The following risks should be mentioned:

– openness of information networks (personal data protection);
– necessity to involve all the students;
– opportunity to influence the process of assessment and rating;
– quantitative rather than qualitative approach to portfolio development.

The main problems in construction of the objective student profile in the form of portfolio are regularity of completion, reliability, objectivity in portfolio assessment.

In addition, portfolio systemizes student activities.

References

1. *The Graduate Student Portfolio – Organize and Energize Your Career Development* URL:http://www.the-aps.org/forum-portfolio.
2. M. Solovyev, I. Abrashkina, D. Kan, Matec Web of Conf. **48**, 06001 (2016)
3. Y. Daneykin, N. Daneikina, Matec Web of Conf. **48**, 06003 (2016)
4. K. Mertins, V. Ivanova, N. Natalinova, M. Alexandrova, Matec Web of Conf. **48**, 06002 (2016)
5. K. Mertins, V. Ivanova, S. Kaftasev, Matec Web of Conf. **91**, 01017 (2016)
6. V. Ivanova, K. Mertins, Matec Web of Conf. **37**, 01025 (2015)
7. K. Cote, T. Emmett, Theory into Prac. **54 I 4**, 352 (2015)

Non-destructive X-Ray testing of complex mechanisms and devices

Ali Ozdiev[1,*], *Yury* Kryuchkov[1], and *Hans-Michael* Kroning[2]

[1]Tomsk Polytechnic University, 634050 Tomsk, Russia
[2]Saarland University, Saarbrücken, D-66123, Germany

Abstract. X-ray tomography becomes a wide-spread non-destructive testing method. Flexibility of the reconstruction algorithms makes possible to apply 3D tomographic analysis to complex mechanisms and systems and detect defects of different types. For instance, this opportunity allows to solve geometrical problem, when the size of the investigating sample exceeding dimensions of the setup. This paper proposes to use non-standard geometry of tomographic scanning and backprojection algorithm optimized for this case to solve previously mentioned problem. Producing and assembling of complex space systems and mechanics requires the testing procedure at each step of the technological process: for separate parts of mechanism as well as for assembled system or device. Presented approach prospectively fits for this task.

1 Introduction

One of the most important advantages of CT measurements is the opportunity to obtain detailed 3D visualization of the sample including its internal structure, which is the way to detect defects: cracks, pores, splits etc. The standard geometry of the tomographic scanning (Figure 1) requires full or limited coverage of the sample by the beam in condition of matching the optical axis of the detector with the center of the sample rotation [1]. Then the standard filtered backprojection algorithm becomes applicable for the reconstruction of the certain part of the sample depending on the type of beam: parallel, fan or cone.

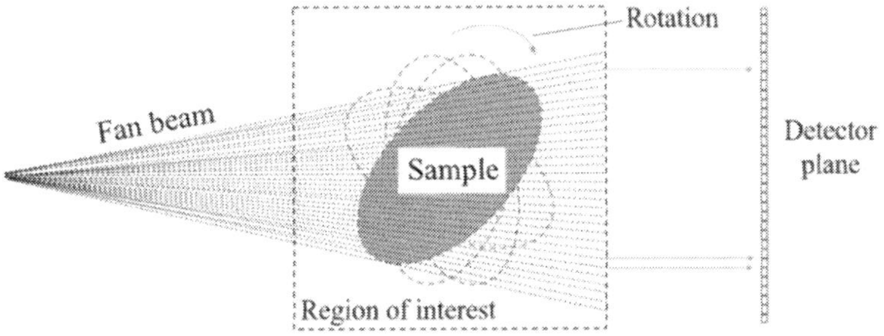

Fig. 1. The standard geometry of tomographic scanning, fan beam case.

* Corresponding author: ozdiev@tpu.ru

© The Authors, published by EDP Sciences. This is an open access article distributed under the terms of the Creative Commons Attribution License 4.0 (http://creativecommons.org/licenses/by/4.0/).

Approaches of the tomographic scanning of the object with sizes, which are bigger, than the dimensions of X-ray setup are still developing. Optimization and adaptation of filtered backprojection algorithms are the way of building the X-ray setup for the scanning of such objects.

1.1 Scanning the objects along their longest dimension

Step-shift scanning was proposed to scan objects with sizes exceeding dimensions of the X-ray setup. Straight forward way to conduct such scanning is to build scanning trajectory along the longest dimension of the sample as it shown in Figure 2.

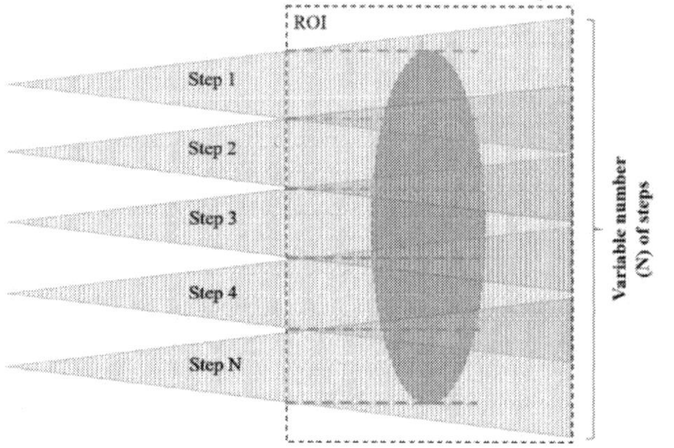

Fig. 2. Step-shift scanning of the sample along its longest view (top view).

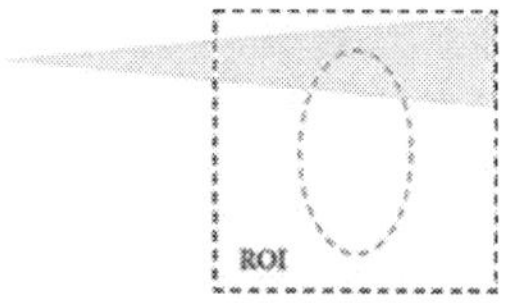

Fig. 3. Backprojection along the exact beam trajectory.

In case when the center of rotation matches with the center of the object one of conditions for applicability of standard backprojection algorithm fails for the entire scanning step except of the only step where the optical axis of the setup matches with the center of the sample. This collision requires the optimization of the backprojection algorithm by ignoring the above condition and projecting the data along the exact trajectories for each scanning step.

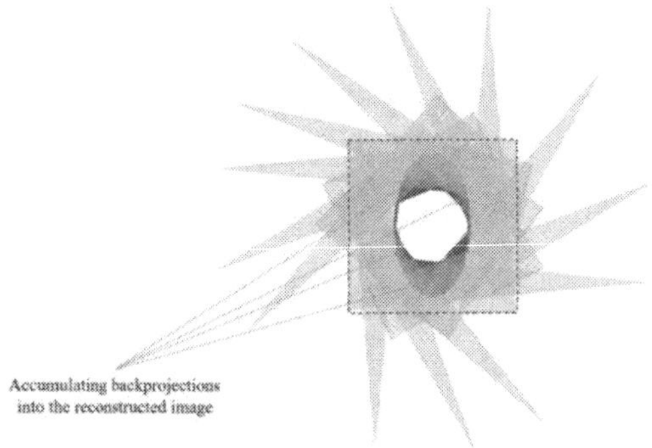

Fig. 4. Reconstruction of the object area from the single scanning step.

Projecting the data along the exact trajectories for each scanning step means that for the certain area of the region of interest, which was covered by the beam at corresponding scanning step (Figure 3), the backprojection for every rotation angle should be obtained.

Interpolation of each value generated by the detector along the real beam trajectories in Cartesian coordinate system mathematically represents the operation of back projection [2]. Reconstructed from the scanned area of the object image is produced by summarizing of all corresponding backprojections (Figure 4).

2 Simulation and reduction of artifacts

Collecting all the reconstructed areas from each scanning step gives opportunity to obtain totally reconstructed image of the object simply by summarizing them to each other. So-called raw reconstructed image (Figure 5), which is the sum of separately reconstructed areas, will contain an artifact – significant intensity difference between internal and external areas of the image.

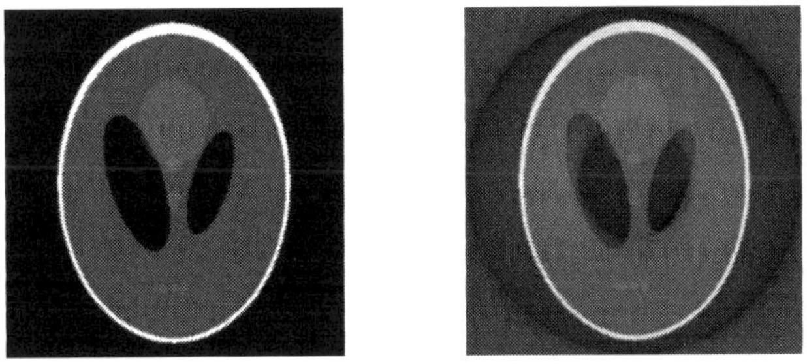

Fig. 5. Original image on the left and reconstructed image on the right with the intensity artefact.

Fig. 6. The sinogram with empty areas and the sinogram where those areas filled with edge values.

Well-known Shepp-Logan phantom (Figure 5, left) was used for the simulation – a standard test image created by Larry Shepp and Benjamin F. Logan for their work in 1974. It usually serves as the model of a human head in the development and testing of image reconstruction algorithms.

The filtered backprojection algorithm was optimized for step-shift scanning, so that a field of view smaller than the object being scanned. A sinogram is a plot of the projection data – horizontal axis is the beam source angle, and vertical axis is the detector number. Because of the limited field of view, certain pixel areas of a sinogram stay empty or simply equal to zero. To reduce the artifact empty pixels of limited view sinogram should be filled out by edge values as it shown in Figure 6 [3].

Fulfilling this condition for every step of scanning leads to the reduction of the artifact and results into the clearly reconstructed image. The reconstructed images from 3 steps of scanning and their sum in compare with the original image are presented in Figure 7.

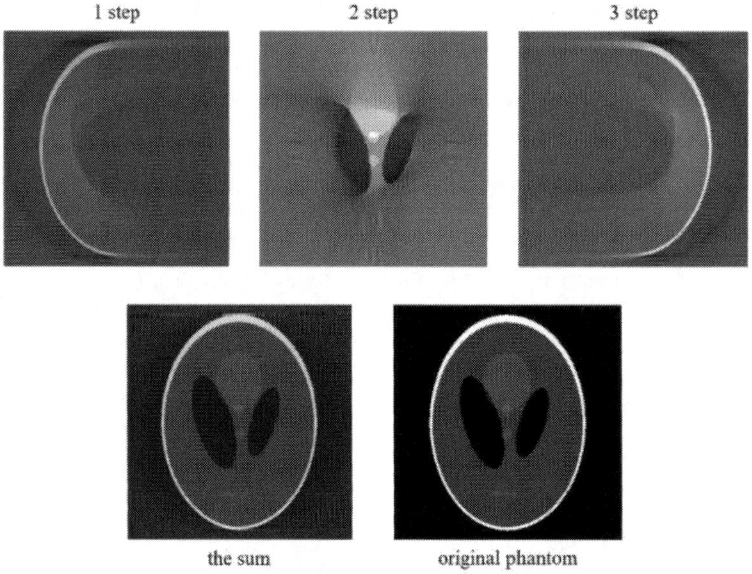

Fig. 7. Top row shows the results of reconstruction from each of 3 steps, second row shows the sum of these steps in compare with original phantom.

3 Conclusions

Optimization of reconstruction algorithm is the prospective way to apply tomographic analysis for complex space systems and mechanisms as one of non-destructive control techniques. Development of new scanning geometries based on such algorithms makes possible to build real-time testing system.

The reported study was partially supported by the Governmental program "Science", research project No. 11.6342.2017/БЧ

References

1. G.T. Herman, *Fundamentals of Computerized Tomography: Image Reconstruction from Projections* (Springer Publishing Company, New York 2009)
2. A.C. Kak, M. Slaney, *Principles of Computerized Tomographic Imaging* (SIAM, Philadelphia 2001)
3. F.E. Boas, D. Fleischmann, JMI **4(2)**, 229 (2012)

The prerequisites of forming a risk management system in the design of facilities space application

Sergey Panychev[1,*], *Vyacheslav* Guzik[1], *Anatoly* Samoylenko[2], and *Andrey* Panychev[3]

[1]Southern Federal University, Department of computing, 347928 Taganrog, Russia
[2]Southern Federal University, Department of radio engineering and telecommunication systems, 347928 Taganrog, Russia
[3]Southern Federal University, Department of antennas and radio transmitters, 347928 Taganrog, Russia

Abstract. The problem of increasing the term of active existence of space-use equipment is relevant. Application of risk management system in the design and manufacture of space-use equipment is a promising approach to increase resiliency and reliability of spacecraft. This paper discusses the preconditions of the risk management system, which is based on the use of critical small amounts of the state of control objects. The technique of statistical processing of the data of the risks based of the additive approximation of the standard statistical distributions is presented. The generalized structure of the adaptive system of statistical diagnostics risk of abnormal conditions in the space application equipment is offered.

1 Introduction

In any project to develop space application objects (SAO) there are many uncertainties. Whenever the process of creating the new system has a significant departure from the usual practice, the result of the development becomes unpredictable. An important task of system engineering is to develop the management system so that the uncertainty was eliminated as early as possible [1]. Any suddenness in the later stages of the development system can cost significantly more than her detection in the early stages.

Problem of assessment and risk mitigation at its core is the task of identifying and eliminating uncertainties at all stages of the SAO lifecycle. Such problems can be solved by analysis, simulation, full-scale tests, which allow to diagnose and quantify the critical important characteristics of the system.

A promising direction for providing an increase fault-tolerance and reliability of SAO is the creation and implementation of an adaptive risk management system for the design and manufacture of SAO. This system will enable detect, locate and remove the causes of defects at all stages of the SAO lifecycle. The testing of system fragments is the means for collecting of important data on the behavior of the entire system and components under controlled conditions.

* Corresponding author: ruu2011@mail.ru

© The Authors, published by EDP Sciences. This is an open access article distributed under the terms of the Creative Commons Attribution License 4.0 (http://creativecommons.org/licenses/by/4.0/).

2 Materials and methods

A detailed description of the risk management is given in [2]. The main disadvantage of used risk management today is to use only quality scale of probability of potential importance of different sources of risk.

The proposed risk management system for the design SAO allows to quantify the significance of risks. It is based on a statistical diagnosis, adaptive to the actual condition of the equipment. To this end, the SAO is introduced into a non-standard mode, which causes random output of basic and additional parameters of the permissible area. Assessment of SAO levels is based on the analysis of level and duration of emissions of monitored parameters.

The risk management system is implemented at three levels. The first level provides a qualitative (rough) assessment. On the second level is produced the approximate continuous quantitative assessment of the actual state. On the third level is made accurate calculation of quantitative characteristics of reliability and adopting risk management solutions.

At the first level of SAO diagnostics is solved the problem of admission control. At this level is given a preliminary assessment of the actual condition of the equipment in the form of "fit – not fit." If any of the monitored parameters has gone beyond the limits of tolerance limits, the risk level is classified as high.

The second level of the SAO diagnostics is implemented, if the result of admission control was the "fit". At this level is produced an approximate quantitative estimate of the risk based on a continuous model of monitored parameters.

At the third level of risk control is made accurate calculation of quantitative characteristics of reliability for inhomogeneous Markov model. It is believed that diagnostic system has properties of adaptability to the structure and the number of available statistics through the use of special algorithms for processing of critical small volume of samples. Deciding on the potentially defective SAO fragments and generation of issue control actions is based on the application of the principles of fuzzy logic and is carried out on all the steps of SAO lifecycle.

3 Results

Figure 1 shows a generalized structure of the adaptive system of statistical diagnostics of risk for abnormal mode of SAO operation with regard to actual state of the equipment [3]. The principle of operation of this system is as follows.

The signals of controlled parameters of SAO (unit 1) $x_i(t)$, $i=\overline{1,N}$ are input to the threshold elements (TE) $2_1 - 2_N$. Here emissions of controlled parameters from tolerance zone are recorded. In unit 3 the parameters with emissions from tolerance zone are detected with an identification of address of the respective channels. In unit 4 the magnitudes of emissions A_{ij}, $i=\overline{1,k}$, $j=\overline{1,N}$ are measured by means of quantization of tolerance ranges to q levels, where k – is a number of parameters, which have emission from tolerance zone, N – is a number of emissions. The value q is determined depending on the specified value of reliability of identification random process $x_i(t)$. Similarly, in unit 5 emission duration τ_{ij} on k parameters through quantization in time of a random process $x_i(t)$ is measured. Accumulated during the process of diagnosing the amplitudes and durations are fixed in the form of emission samples formed respectively in units 6 and 7.

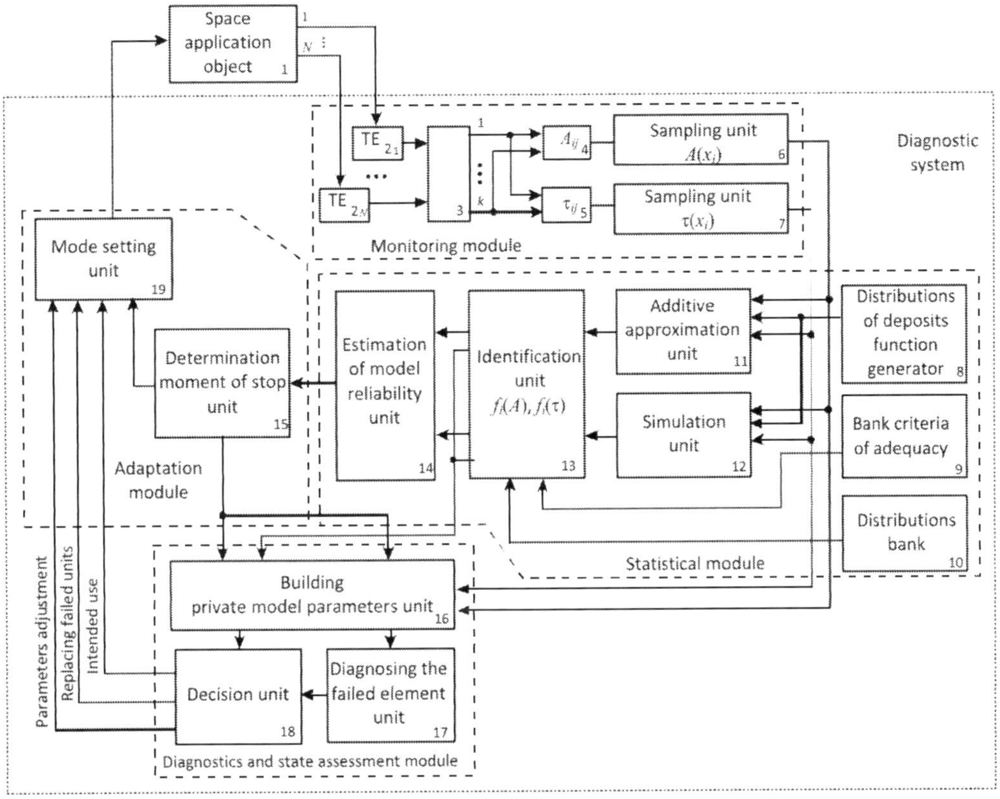

Fig. 1. Generalized structure of the adaptive system of statistical diagnostics of risk for abnormal mode of object operation with regard to the actual state of the equipment.

The sampling of amplitudes and durations parallel is given to the units 11 and 12. Unit 11 implements a method of additives for constructing the empirical density function of the distribution. Unit 12 implements the algorithm of kernel estimation of empirical data. Inputs units 11 and 12 are also connected to the generator output distributions of the function of deposits necessary for the implementation of method of processing of arrays of small samples. In unit 13 an identification of functions of amplitude distributions and emission durations of SAO diagnostic parameters $f_i(A)$, $f_i(\tau)$, which have been received in units 11 and 12, are performed. This procedure is made by sequential revision a plurality of the most common laws of probable distributions (equiprobable, exponential, normal, Rayleigh, Weibull et al.) from the bank of distributions 10. Calculations of corresponding values the criterion of consent are performed using the bank of criteria for adequacy 9. The designing of ranked series, based on which a decision on the conformity of empirical data to one of the theoretical distributions, is performed in unit 13.

The validity of the estimated statistical models of the individual parameters is assessed in unit 14, according to results of unit 13. Doing it, we calculate the conditional density distribution emissions of each controlled parameter $f_i(E) = f_i(A) f_i(\tau / A) = f_i(\tau) f_i(A/\tau)$.

In accordance with the theory of reliability the emission distribution density is equal to the frequency of failures i-th parameter. Then, the reliability (probability of failure-free operation) of the test piece for the i-th parameter is estimated by the formula

$$P_i(t) = 1 - \int_0^t f_i(E) dx . \qquad (1)$$

Equation (1) gives a quantitative description of the significance of different sources of risk.

In unit 15 the proposition about stopping control process or its continuation, based on the calculation and analysis of the trajectory of the functional loss and the terminal benefit, is decided [4]. If unit 15 detects the inadequate reliability of the statistical models of the some parameters, then the unit of mode setting 19 generates a signal to continue the monitoring process. If unit 15 decides to move to the statistical diagnostic mode, then the unit of diagnostics and assessment of the SAO state will be initiated.

In unit of the construction of private models of parameters 16 is carried out formation of continuous models of efficiency for individual SAO parameters on the basis of sampling small size and probability models in form of two-dimensional functions of the density distribution for emission characteristics . Thereafter the intensities of gradual failures are calculated and an inhomogeneous Markov models for the three statuses are constructed. Based on result of operation of unit 16 the diagnosing of the failed element of SAO is performed in unit 17 that implements a technique localizing faults in the face of uncertainty.

Unit 18 uses the results of units 16 and 17 for producing three types of the control action to unit 19:

- the use of the SAO for its intended purpose in normal operation;
- regulation of unstable parameters and the SAO output to a normal operating mode;
- replacement of the failed unit (or fragments) of the SAO.

4 Discussion

The full-scale simulation and test of object fragments (and subsequently the facility as a whole) is applied based on the threshold principle of formation of the parameters emissions from the acceptable zone. As a result, monitoring values are presented as a series of random of levels of emissions object parameters [3].

Statistical processing of emissions critically small volumes is based on the additive principle. The emissions are presented as a set of standard symmetric distributions (deposits), or as a set of arrays of random values generated approximately the point of emission parameter.

Each mode of full-scale simulation is displayed as the numerical statistical characteristics of the corresponding random processes.

5 Conclusions

Thus, the proposed statistical diagnostics risk system can be put in the basis of full-scale tests and simulation of SAO fragments. This will mitigate the risks associated with the design and reduce the cost of resources in the production of facility.

References

1. A. Kossiakoff, W.N. Sweet, S. Seymour, S.M. Biemer, *Systems Engineering Principles and Practice* (A John Wiley & Sons, INC. Publication, 2011)
2. L. Wilson, D. McCutcheon, *Industrial Safety and Risk Management* (Edmonton, University of Alberta Press, 2003)
3. V.F. Guzik, V.I. Kidalov, A.P. Samoylenko, *Statistical diagnostics of non-equilibrium objects* (SPb, Sudostroenie, 2009)
4. A.P. Samoilenko, A.I. Panychev, S.A. Panychev, *SIBCON*, 7491722 (2016)

Wide dynamic range 500 fA sensitivity current measurement instrument

Anatoly Pelemeshko[1,*], *Aleksey* Styuf[1], *Vitaliy* Prokopyev[1], *Yuri* Prokopyev[1], and *Alexander* Zadorozhny[1]

[1]Novosibirsk State University, 630090 Novosibirsk, Russia

Abstract. Main technical features of Low Energy Charged Particle Sensor (LECPS) are described, taking into account LECPS high sensitivity, wide dynamic range, and conditions of in-flight operation on-board a satellite. LECPS input cascade is capable of current measurement as low as 1 pA within a 120 dB dynamic range, providing error level below 20%. Statistical calculations of acquired data, measuring time considerations and implementation of input cascade auto-calibration technique were applied to gain the best possible stability and accuracy over –20° to +50°C operation temperature.

1 Introduction

Division for Atmospheric Research at Novosibirsk State University has been designing on-board space devices for the last 35 years. Most activities are devoted to automation of space experiment by creating measuring equipment for inflight operation that meet strict requirements of space environment, high reliability, mass and power consumption constraints. Another scope of the laboratory is telemetry and interface devices that simplify connection of external probes to power and information interfaces of a satellite.

Low energy charged particle flux has always been a matter of concern for spacecraft engineers since it inevitably leads to satellite charging and results in an uncontrolled electrostatic discharge [1]. Main sources of particle flux are atmospheric plasma, solar particle flux and plasma thrusters [2]. Either a satellite is sent to a new orbit or new design/materials are implemented within a satellite it is necessary to investigate a charge-discharge phenomena by providing in-flight experimental data.

An Low Energy Charged Particle Sensor (LECPS) was designed at the Division for Atmospheric Research to allow in-flight monitoring of low energy charged particle flux. LECPS photo and specifications are presented on Figure 1.

LECPS is meant to be placed on the outer side of a satellite. According to the existing space environment model and industry requirements, the device must register current within 1 pA to 100 nA range. For this reason, LECPS is equipped with a wide dynamic range current meter obtaining a 500 fA sensitivity and operating in a temperature range from –20° to +50°C ensuring an error level below 20%. A light weight, small dimensions and low power consumption of the device combined with a 15 year lifetime provides

*Corresponding author: pelemeshko55@gmail.com

© The Authors, published by EDP Sciences. This is an open access article distributed under the terms of the Creative Commons Attribution License 4.0 (http://creativecommons.org/licenses/by/4.0/).

possibilities for integration into nearly every up to date space mission. LECPS information interface is compatible with a widespread MIL-STD-1553B [3]. Compliance with all stated requirements was confirmed by accurate ground tests.

sensitivity:	500 fA
dynamic range:	1 pA – 100 nA (120 dB)
error level:	20%
dimensions:	70 × 70 × 54,5 mm^3
weight:	0,47 kg
power consumption:	1,5 W
temperature range:	–20° to +50°C
lifetime:	15 years
interface:	MIL-STD-1553B

Fig. 1. Low Energy Charged Particle Sensor (LECPS) photo and specifications.

2 Solutions

Figure 2 presents functional diagram of the LECPS. An input cascade consists of a low-input current pre-amplifier as a current-to-voltage converter and a programmable gain amplifier (PGA). As the result, an input signal has a leakage current less than 100 fA, while output signal can be multiplied by factor of 1, 10, 100 and 1000. PGA output signal drives an ADC (12 bit, 200 Hz) that is integrated into a microcontroller (MCU) periphery. MCU performs data acquisition, controls a gain value of PGA, and interfaces to an external MIL-STD-1553B command bus. An input relay allows measuring of "zero-level signal" at all gains thus performing online calibration to get rid of input cascade offset caused by temperature drift and parts aging.

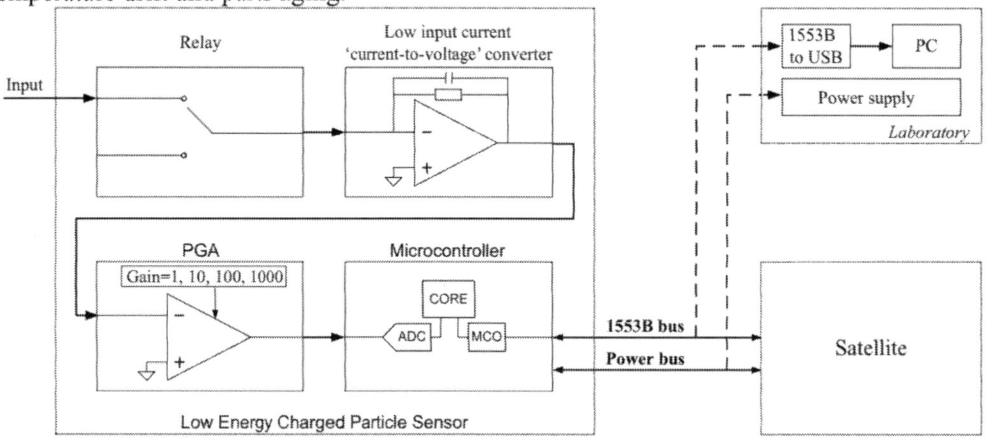

Fig. 2. The functional scheme of LECPS: current meter input cascade and data acquisition.

To obtain acceptable ADC data array for further calculations the following requirements for electronic components should be considered:

- input operational amplifier (in a whole temperature range):
  - input bias current below 100 fA;
  - offset voltage below 100 uV;
  - low temperature drift less than 10 uV/°C
  - unity gain below 1 MHz;
- PGA (in a whole temperature range):

- digitally programmable gain;
- selectable gains 1, 10, 100, 1000;
- low temperature drift (referred to input) less than 10 uV/°C

- ADC (integrated into microcontroller):
  - sufficient resolution – 12 bit;
  - data acquisition rate – 10 KSPS or higher.

Time chart of a measuring period is presented in Figure 3. This procedure consists of auto-calibration phase (relay off) and current value acquisition phase (relay on). During each phase the MCU successively switch gain value of a PGA and calculates a corresponding average value.

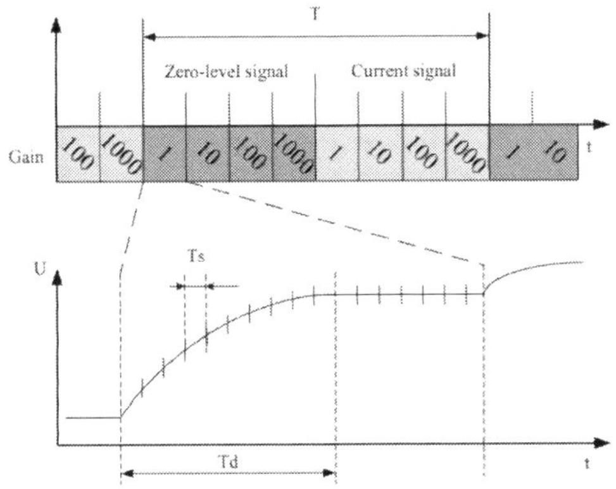

Fig. 3. Survey timing diagram of LECPS measuring cascade: T – single measurement time, Ts – ADC time survey, Td – "dead time".

Obviously, both a relay switch and a gain change cause transients on the output of input cascade. Hence, MCU skips ADC values during a defined "dead time" Td to avoid transients effecting calculations. Nevertheless, the ADC unit is still kept operational to maintain electrical circuitry at steady state. The estimated Td time is 100 ms for gain values 100 and 1000, and is negligibly low for gain values 1 and 10.

A calculated average value is transferred to a host device via a MIL-STD-1553B compliant interface, supported by MCU due to an integrated peripheral unit. During in-flight operation LECPS is linked to MIL-STD-1553B command bus of a satellite on board computer. For the purpose of ground tests a standard 1553B-to-USB converter offers connection to any laboratory PC or laptop (see Figure 2). In this case all measured data is visualized in a specially designed software monitor.

3 Results

A standard deviation of a resulting current value was studied as a function of number of ADC values (e.g. storage time). This was done to ensure sufficiency of electrical filters at schematic level. The result is shown in Figure 4. An evident $\frac{1}{\sqrt{n}}$ dependency demonstrates that ADC input signal has a "white noise" nature. A 10 sec storage time was used during future calibration procedure to provide a standard deviation below 20% at low input current values.Calibration of LECPS in the completely input current range from 1 pA to 100 nA was carried out using Keithley SourceMeter 2611B. A special triaxial cable technique and a metal shield was used to avoid electromagnetic disturbances affecting signal at input current below 50 pA.

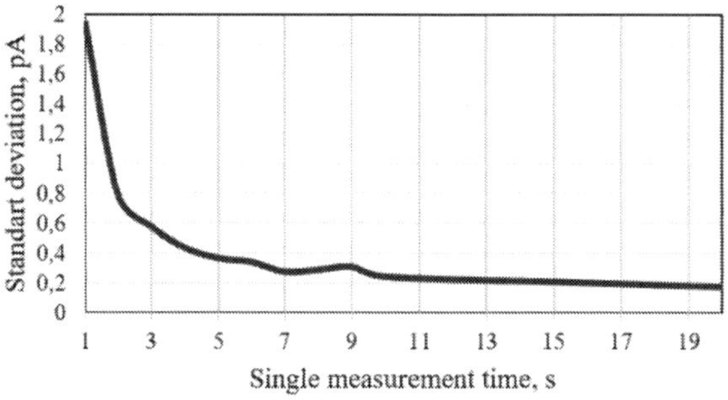

Fig. 4. The dependence of the standard deviation on measurement time.

Temperature stability of the LECPS was investigated during tests in a climatic chamber in an extended temperature range from –55° to +75°C. Figure 5 exhibits advantage of auto-calibration implementation. As the result, auto-calibration technique leads to a reduction of the calculated current value change from 10 000 to 100 ADC quants.

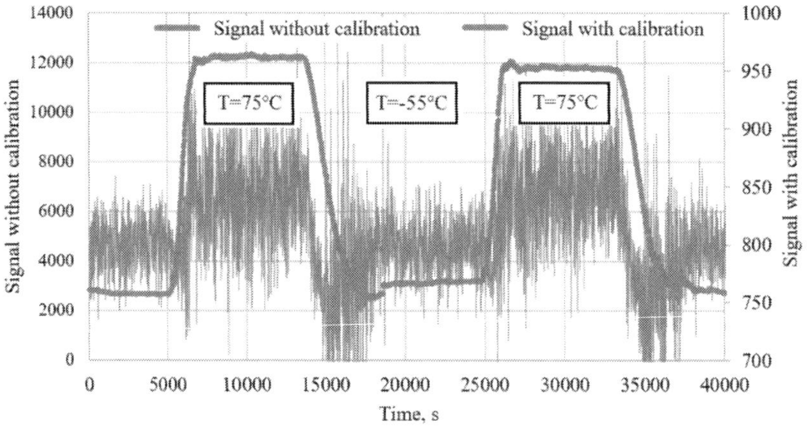

Fig. 5. The measurements at temperature -55°C and +75°C.

4 Conclusion

A complete design cycle was carried out to create the LECPS – wide dynamic range 500 fA sensitivity current measurement instrument. Precise ground tests proved LECPS temperature stability and metrological characteristics, as well as ability to be operated on board a satellite.

References

1. L. Aristov, M. S. Borodin, et al, *Nauchno-tehnicheskie razrabotki KB "Saljut"* **3**, 327, (2011) (in Russian)
2. S. V. Balashov, V. V. Ivanov, et al, *Reshetnevskie chtenija* 46, (2005) (in Russian)
3. *Condor Engineering, Inc, MIL-STD-1553 Tutorial (1600100-0028)*, (2000)

Internal education quality evaluation as a factor of development of engineering education quality in Russia

Antoniy Shvindt[1], and *Ivan* Nikanorov[1,*]

[1]Education Quality commission of Youth affairs Council of Ministry of Education and Science of the Russian Federation, 115184 Moscow, Russian Federation

Abstract. In the paper, general approaches to evaluation of quality of engineering education in Russia, stated by expert community, are considered. The authors arrived at the conclusion that attention to internal evaluation mechanism of the education quality in Russian normative framework of higher education system is insufficient, as compared to European higher education practices. A pattern of involving the students in the internal evaluation of higher education quality is submitted to be a tool to develop the quality of engineering education. The pattern is been piloting in the framework of All-Russian project "Involving of Students of Higher Educational Institutions in Evaluation and Development of Education Quality "High-Quality Education".

1 Introduction

Development of higher engineering education quality was repeatedly declared to be a top target of Russian higher education general development. It is asserted that without that, to obtain competitive ability and technological sovereignty is impossible [1]. Herewith, it is to be noted that engineering staff training in Russia, as well as that in other education sectors, has been undergoing thorough reforms for the latest 25 years. That has resulted in a set of common negative trends pointed by experts, employers and students themselves: poor standards of enrolees, poor material and technological resources to maintain the programmes, and fails in adequacy of moulded acquirements to employers' requirements (although the quality of the engineering staff training is considered to differ [2]).

In a number of comparative investigations, the challenged processes in engineering education are noticed not only in Russia, but also in other countries [3, 4]. Particular attention is given to correlation of the enrolees' acquirements and educational progress up to finishing higher education [5]. An exploration is worth noticing, concerned to the quality of engineering education in BRICS countries, exposing a large disparity between the quality of elite and massive engineering educational programmes there. At the same time, the total number of engineering programmes graduates is considered to be a determinative

* Corresponding author: nikanorov@zaobrazovanie.ru

© The Authors, published by EDP Sciences. This is an open access article distributed under the terms of the Creative Commons Attribution License 4.0 (http://creativecommons.org/licenses/by/4.0/).

number in the global labour-market of high-technology spheres of economy [6]. It is to be noted that the content of the curricula and their quality management receive little attention.

2 Analysis of Russian and international background

In this context, it is critical to underscore that in Russian Federation, the education quality evaluation is particularly exposed with external tools, which is state system of education quality control above all – licensing, accreditation, independent quality assessment (also triggered and funded by state government authorities). As for professional and public accreditation and attainment certification, if they are provided by normative framework of higher education system and have had a definite success [7, 8], they are not fully implemented and still lie ahead for next ten years.

Russian Federation's mainstreaming into European Higher Education Area and signing the European Standards and guidelines for quality assurance in the European Higher Education Area (ESG) in 2015 has given a fresh impetus to the higher education development in Russian Federation. Standards related to internal evaluation of the education quality (eg, 1.3, 1.6, 1.9) provide the following principles: (1) "active role in creating the learning process", (2) necessity of sufficient "funding for learning and teaching activities" and accessibility of learning resources, (3) urgent need to carry monitoring continuously and to review degree programmes to make them correspond "to the needs of students and society", and need to communicate on the improvements "to all those concerned", and (4) the unity of the internal and the external education quality systems.

The implementation of the ESG student-centred principles noted above is impossible without formation of student communities aimed to evaluate and develop the education quality (including diversification of possible educational paths and development of various modes and ways to obtain formal, as well as informal education), to widen the students' estimation tools, and to form the mechanisms of incorporating this estimation into both external and internal education quality evaluation.

To speak about approaches to providing the internal assurance of the education quality in Russia, it is worth noting G. V. Andrushchak's paper on existing practices on the back of international ones [9]. It appears that Russian educational institutions, compared to Western ones, firstly, severely constricts a students' choice of educational path, relevant to their future career; secondly, lack an efficient practice of students' rating the teaching work in order to improve it [9]. Therewith, it is acknowledged that the essential factor of the efficiency of the mechanisms is proper motivation of the student community.

This paper was published before enactment of the Federal Law N 273 "Education in Russian Federation Act", which makes allowance for bodies that have a right to participate in the institution administration through representing a position of a group of educational relations actors. Thus, the students' participation is possible in case of their involving in governing bodies of educational institutions (Boards of Studies, supervisory boards, etc, including opportunities provided by institution Charter). Among the tools of involving the students in evaluation and development of education quality, an important one is procedure of self-examination of an educational institution, captured by section 29 of the Law.

It is also to be pointed out that, as distinct from the ESG, the "Education in Russian Federation Act" does not capture the internal evaluation of education quality, and this estimation refers to the competence of the educational institution itself.

3 The submitted pattern

With reference to the above mentioned, the Education Quality Board of Youth Council of Ministry of Education and Science of the Russian Federation (hereafter – the Board) considers development of student self-government to be a necessary tool of internal evaluation of education quality. The Board submitted a conception of an on-going student union, with its activity being aimed to evaluate and develop the education quality (student board). An All-Russian youth project "Involving of Students of Higher Educational Institutions in Evaluation and Development of Education Quality High-Quality Education" (hereafter – the Project) has been triggered by the Board. The Project model is focused on synergistic interaction of the following actors: higher educational institution, public interest organisation, federal executive authority responsible for elaborating the state policy in the education sector or youth policy. The goal is to develop youth self-governance in educational institutions and to engage public interest organisations in continuous monitoring and control of education quality. The Project is implemented in a concerted effort by Ministry of Education and Science of the Russian Federation, the Education Quality commission of Youth affairs Council of the Ministry, and All-Russian public interest organisation "For the Quality of Education".

More than 6.000 students of 208 higher educational institutions from 70 member states of the Russian Federation are involved in the implementation of the Project. Apart from core activity of the unions, the member students are continuously being engaged in e-learning (in MOOK format), and in exchanging "the best practices" in project activities, which enabled to form a social inter-academic student network united with common tasks and involved in project activities. The scope of tools using by them embraces the following.

(1) Integrated monitoring allows estimation the students' satisfaction of infrastructure, conditions of learning process functioning (equipment etc), internal learning process management (timetable, logistics etc), and general satisfaction of teaching staff. To carry the monitoring, a complex questionnaire is made (based on student standard) to be used once a semester.

(2) The scope of the second tool is limited by a single subject studying as a part of a degree programme. Subject monitoring, being carried at the end of each lecture / seminar, helps to evaluate clearness of lectures, a teacher's punctuality, social skills, using ICT, adequacy of studies content to the subject matter, orientation to declared acquirements to be obtained. Processing of the results can be used for making a student ranking of teachers, establishing the students' feedback and improving proposals for the teaching.

(3) Annual self-examination monitoring estimates the students' awareness of their own place in the institution. The results can be used to raise the purposefulness of the student engagement, to realise their expectations, and to establish the scope of information support sector. The questions touch on reasons the students have entered the institution, have chosen it and their special field, their career plans, subjects they want to study, learning results expected to be obtained. Also, their familiarity with Federal and local documents on their field of education and the regularity of preparations for studies is inquired.

(4) General monitoring results in observing general trends and dynamics of oscillations of the students' feelings about the learning process in the real-time mode. It should be carried on a regular basis – weekly or even daily – and contain the following questions: were the former day / week useful or not, have they learned anything new for this day / week, have they succeed to prepare for all the studies?

(5) With longitudinal explorations, changes of the same group of students' satisfaction are estimated for a definite period of time. Questions about changes of attitude to the necessity of disciplines, to activities of teaching staff, to practical trainings, being asked several times for extended period, allow creating more or less non-biased image of a

discipline and its teaching. The survey should be conducted three times: in current semester: a semester later; after completing an externship or 3 to 6 months after employment in the profession during the upper years of studying of after graduation.

Thus, in the framework of the submitted pattern, a student community is considered to be an equal partner to the administration of the educational institution. The work product of the community can be used in a set of management procedures in the institutions.

4 Conclusions

The challenges of engineering education in Russia pointed by experts – comparably lower attainment level of enrolees, problems in material and technological resources maintaining the programmes, and the necessity to review the content of the curricula – actually seem to exist. As a solution, implementation of the mechanisms of internal evaluation along with external ones is suggested. Efficient involving of the students to the evaluation of the education quality would allow improving the competitive ability of the degree programmes and the efficiency of executive decision-making. That would make the programmes more flexible and oriented to employers and future graduates' interests.

References

1. *Stenogramma zasedaniya soveta pri Prezidente Rossijskoj Federacii po obrazovaniyu i nauke ot 23 iyunya 2014 goda.* URL: http://kremlin.ru/events/president/news/45962 (2017) (In Russian)
2. L.M. Ogorodova, V.M. Kress, Yu.P. Pokholkov, Engineering Education **11**, 2. (2012)
3. E. Kardanova, P. Loyalka, I. Chirikov, L. Liu, G. Li, H. Wang, E. Enchikova, H. Shi, N. Johnson, Assess Eval High Educ **41**, 5 (2016)
4. P. Loyalka, M. Carnoy, I. Froumin, R. Dossani, J. B. Tilak, M. Dobryakova, Higher Education **68**(6) (2014)
5. A. Light, W. Strayer, JHR **35**, 2 (2000)
6. P. Loyalka, M. Carnoy, I. Froumin, R. Dossani, J. B. Tilak, *The Quality of Engineering Education in the BRIC Countries* (2013), URL: https://reap.fsi.stanford.edu/sites/default/files/Quality_of_engineering_education_in_the_BRIC_countries.pdf
7. *Kriterii i procedura professional'no-obshchestvennoj akkreditacii obrazovatel'nyh programm po tekhnicheskim napravleniyam i special'nostyam: informacionnoe izdanie* (TPU 2014) (In Russian)
8 V. Ivanova, K. Mertins , S. Kaftasev, MATEC Web of Conferences **91**, 01017 (2017)
9 G. V. Andrushchak, Voprosy ekonomiki, **6** (2007) (In Russian)

X-ray tomography of the aerospace products

Vladimir Smolyanskiy[1,*], *Maxim* Rychkov[1], and *Valeriy* Borikov[1]

[1]National Research Tomsk Polytechnic University, 634050 Tomsk, Russia

Abstract. Currently, non-destructive methods are increasingly being used to inspect the critical components in different branches of industry. Special focus is on the usage of high-energy X-ray microtomography used to inspect large objects with high density. This method provides high accuracy and a possibility to measure the internal and external dimensions of the test sample without its destruction. It allows obtaining the information about internal defects and damages as well as manufacturing and assembling quality. This article describes a tomographic study of a medium-size high-density sample with a resolution about of 70 microns.

1 Introduction

The quality control of aerospace products is one of the steps in their manufacturing process. During this phase, operational safety and product reliability are evaluated.

Despite the variety of known methods [1-2], the study of high-density objects remains an unsolved problem, because product thickness and dimensions prevent from carrying out an inspection without sample destruction. A solution came with the discovery of X-ray testing and computed tomography (CT) [3].

The main objective is to detect defects in the internal ferrite rings of a rotor position sensor with a diameter of 3 mm and to measure the depth of a 1 [mm] weld. Reliability and failure-free service life of the electro motor depend on manufacturing and assemble quality control. It requires inspection of the motor assembly, which is only possible with X-ray tomography.

2 Materials and methods

The object of the study is the motor DBE63 designed for operation in electric pump (LZTK-2) in the thermal control system of a spacecraft with on-orbit operating lifetime of about 15 years. The motor is 90 mm in diameter and consists of titanium and aluminium parts with a thickness ranging from 6 to 12 mm (Figure. 1).

The inspection is performed on a cone-beam computed tomography scanner (CBCT-scanner) for large objects designed by Tomsk Polytechnic University (TPU). The CBCT-scanner consists of an X-ray source MXR-451HP/11 (COMET), a flat panel detector XRD-1642 (PerkinElmer) and a linear rotary table DCNI 50 ST (DELKEN).

* Corresponding author: vsmol@tpu.ru

© The Authors, published by EDP Sciences. This is an open access article distributed under the terms of the Creative Commons Attribution License 4.0 (http://creativecommons.org/licenses/by/4.0/).

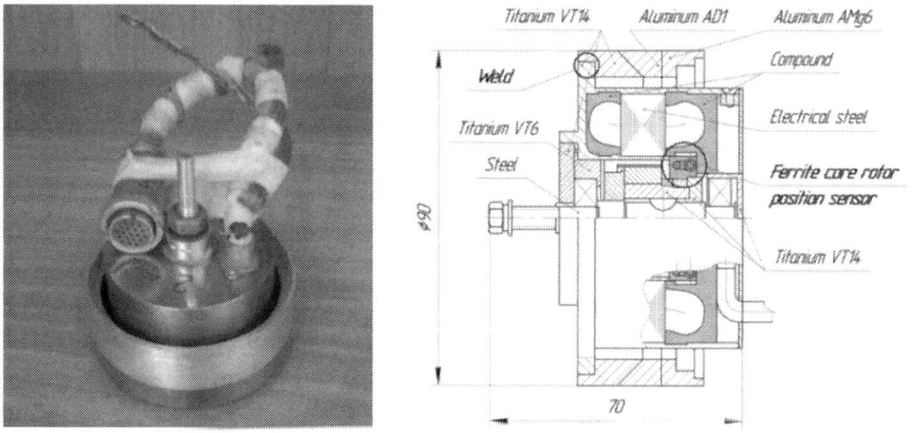

Fig. 1. Electro motor DBE63.

The cone-beam method provides high rendering performance and allows varying the sample dimensions and spatial resolution [3] (Figure. 2).

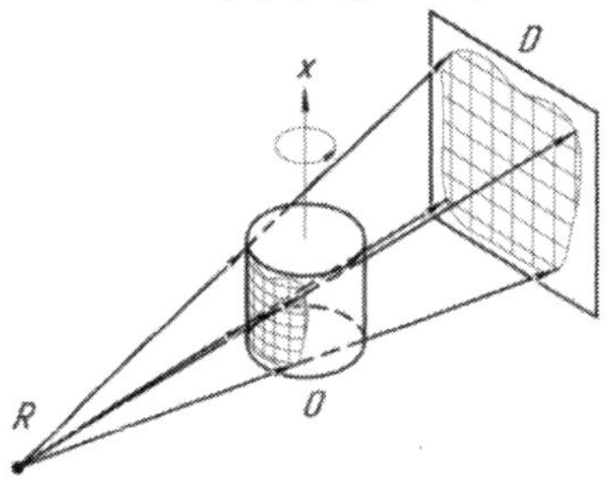

Fig. 2. Schematic diagram of the cone-beam method Legend: R – X-ray source, O – test sample, D – detector, x – rotation axis.

According to the thickness of the inspected material, the scan mode is set to: 300 kV accelerating voltage, 2.3 mA anode current, 0.3 deg projection angle increment, 1 s exposure period, 720 projections, 45 min data collection time.

The INKCT software package (TPU) collects shadow projections. The NRecon software package (Skyscan) performs the 3D reconstruction (Figure. 3.a) by filtered back projection [5, 6]. Finally, the CTvox software package (Skyscan) implements further volumetric visualization of the reconstructed slices (Figure.3.b).

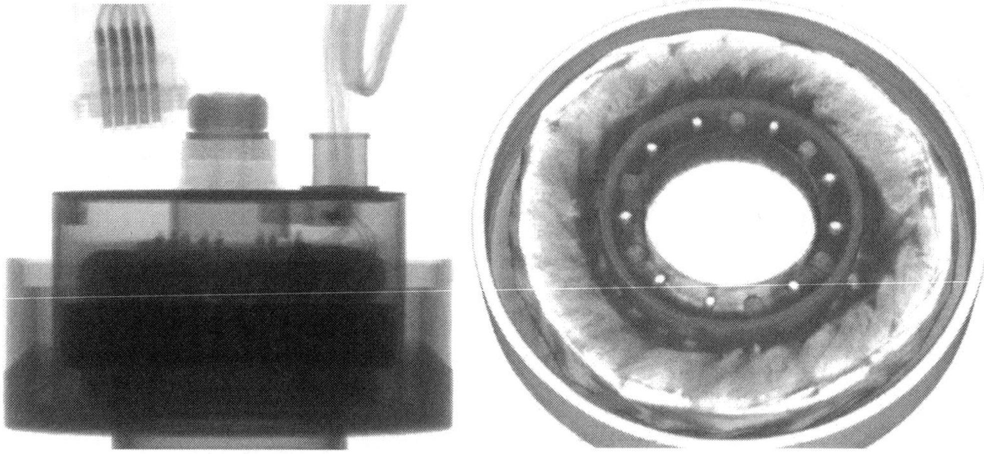

Fig. 3. a) Motor shadow projection, b) Motor cross-section near the ferrite rings.

3 Results

Computed tomography, reconstruction, segmentation and visualisation are performed to study the object in detail. In addition to the outer surface of the sample, the model shows its internal elements in accordance with their density. Cross-section images of the complex internal structure bring hidden defects into clearer focus without any destructions of the electro motor. The analysis results show the position of ferrite rings, their dimensions (diameter of 2.986 mm) and clear boundaries. The inspection reveals the cause of the trouble in motor operation. There is one of the ferrite rings with a through-thickness crack (Figure. 4).

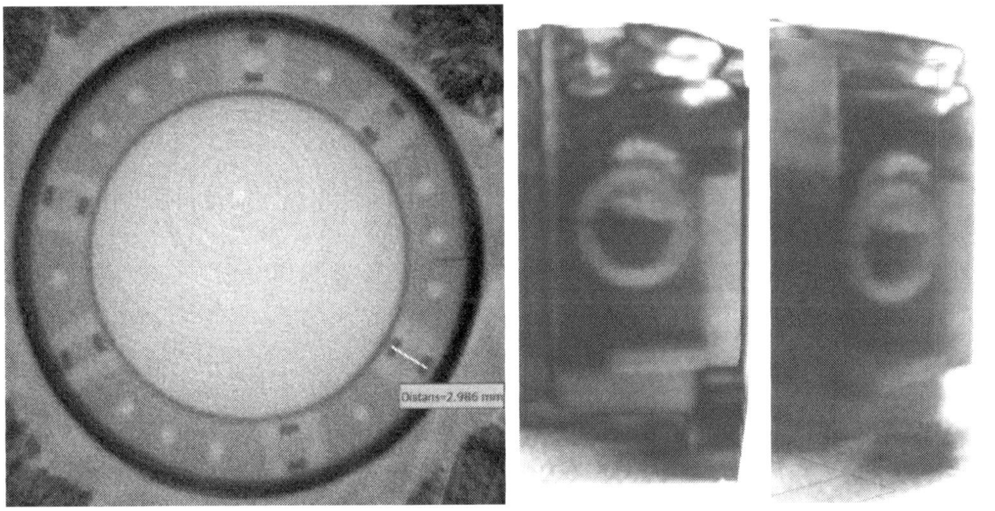

Fig. 4. Through-thickness cracked ferrite ring.

The internal structure of the object was reconstructed to inspect the weld of the motor cover. The images were obtained at the X-ray tube voltage 225, 250, 300, 350 kV and current 3.0, 2.5, 2.3, 2.0 mA, respectively. We did not find the weld on the reconstructed slice (Figure 5) by comparing the detail drawing and the object shadow projection.

Fig. 5. The object shadow projection and the reconstructed slice with the weld.

4 Discussion

For the given voltage and current, the weld is not defined. This is caused by the low penetrating power of radiation and the lack of differences in density (weld placed between titanium alloys VT14).

Since the base material of the construction is titanium (a high-density material) and the size of the weld is 1 mm, the CBCT-scanner must use a more powerful ionizing radiation source with harder rays and higher penetrating power, for example, with a betatron MIB-4 (TPU) and linear array detector IHE2 (Detection Technology). It is necessary to provide an X-ray beam hardening by special copper filters that absorb soft radiation. It would increase the average power and improve the reconstruction quality. An X-ray beam must be collimated by lead blocks to reduce the scattered beam influence and to provide a high spatial resolution. The scan settings for a new installation should be approximately as follows: accelerating voltage – 500 kV, projection angle increment – 0.3 deg, exposure period – 3-5 s.

5 Conclusions

CT is a powerful tool to study the internal structure of an object. In this article, we describe the tomographic scanning of the space-worthy electro motor DBE63 as well as its following reconstruction and analysis. This experiment allows us to inspect a medium-size high-density sample with a resolution of about 70 microns and to detect product defects affecting its performance.

However, further study will focus on the possibility of using high-energy high-resolution X-ray microtomography for the inspection of a large-size high-density object. A promising new application for the future is using of a betatron will be as a more powerful ionizing radiation source with harder rays and higher penetrating power. This work shows the importance of X-ray tomography for non-destructive testing in the space industry. The results of this work demonstrate the efficiency of the developed CT system.

References

1. V. P. Vavilov, D. D. Burle, NDT&E Int. **73**, 28 (2015)
2. V. N. Borikov, O. V. Galtseva, G. A. Filippov, J. Phys.: Conf. Ser. **671**, 012014 (2016)
3. S. V. Chakhlov, S. P. Osipov, A. K. Temnik, V. A. Udod, Rus. J. of Nondestructive Testing **52**, 235 (2016)
4. T.M. Buzug, *Computed Tomography: From Photon Statistics to Modern Cone-beam CT* (Springer Science amp; Business Media, Berlin, 2008)
5. B. Wolkowski, E. Snead, M. Wesolowski, J. Singh, M. Pettitt, R. Chibbar, S. Melli and J. Montgomery, J. of Synchrotron Radiation **22**, 1130 (2015)
6. Feldkamp L. A., Davis L. C., Kress J. W., JOSA A. **1**, 612 (1984)

MATEC Web of Conferences **102**, 01034 (2017) DOI: 10.1051/matecconf/201710201034
V International Forum for Young Scientists "Space Engineering"

The model of specialists training system for high-tech industries

Mikhail Solovyev[1,*], *Kseniya* Mertins[1], and *Anton* Shagdyrov[1]

[1]National Research Tomsk Polytechnic University, 634050 Tomsk, Russia

Abstract. The article reviews the world experience of the specialist training model for high-tech industries. The focus is made on the approach to organization of training to help students develop enterprise-technological skills implemented in National Research Tomsk Polytechnic University. The model considers the strategy of scientific and technological development adopted in the Russian Federation and the need of high-tech industries to improve the quality of educational and training programs and training of the personnel to meet the demand of today's environment who are ready to face future challenges and to establish new markets, and creates the conditions for development of entrepreneurial and technological competences.

1 Introduction

In the rapidly changing world of technology, the key determinants of the enterprise competitiveness are ongoing efforts to generate new knowledge and readiness for introduction of technological and organizational innovations (rapid changes in technology and financial chains, production cycle restructuring, etc.). This puts forward new requirements for staffing of industries. This is the development trend of modern world universities. Research universities are gradually transformed into (Univ. 2.0) entrepreneurial universities (Univ. 3.0), and then into universities that create new markets and build market infrastructures [1] (Univ. 4.0). In the «Strategy of scientific and technological development» of the Russian Federation it is required introduction of technologies, products and services to the modern market: digital manufacturing technologies, robotic systems, new materials and construction methods, big data processing, clean energy, and others [2]. The Strategy obliges the universities to transform into open educational and scientific-technological platforms for development of engineering and training of tech entrepreneurs.

2 Requirements of modern high-tech industries

There are three main types of specialists in the field of engineering: engineers who operate technological equipment; engineers who implement technology and development; engineers who develop new technologies and equipment.

*Corresponding author: solo@tpu.ru

© The Authors, published by EDP Sciences. This is an open access article distributed under the terms of the Creative Commons Attribution License 4.0 (http://creativecommons.org/licenses/by/4.0/).

The research identified a number of basic requirements for future employees of high-tech industries, the engineers of all the three types [3]:

– profound scientific, mathematical and engineering knowledge;

– critical awareness of the forefront in the professional field, the ability to apply new and emerging technologies in the field of specialization;

– innovative approach to develop new original ideas and solutions of engineering problems (to be able to identify, properly formulate and solve practical engineering problems, including implementation of digital modeling as the basis for design and engineering); detailed understanding of the scientific principles of professional activity, the ability to plan and carry out research;

– an ability to plan the product life cycle and willingness to manage the project from concept development on down to launch on the market.

Many years' experience and a variety of approaches to training of specialists in the leading Russian universities allow effective preparation of specialists to meet these requirements [4].

A new market demands sharp reduction in the time between knowledge acquisition and creation of technologies based on this knowledge pose additional and critical requirements that apply to employees of high-tech companies: critical thinking, emotional intelligence, change management, the ability to manage people and interact with them, teamwork, etc. [5]. These competences are crucially important for of specialists of 2nd and 3rd types. The educational programs of universities must respond to changes in these requirements.

3 Experience in training engineers of today

Moscow Polytechnic University. Large-scale implementation of new educational programs and technologies: STEM education enabling to help students bridge the gap between educational objectives and activities of a scientist and an engineer. Master's degree program in production-engineering (programs are based on promising R&D-projects concerning topical areas of engineering and IT science, implemented to the order or in cooperation with leading IT companies [6], in accordance with STI [7] and Horizon 2020, and others. A compulsory course "Project Activities" involves 4 to 12 people, specialising in different areas (1st year), within which various engineering projects are performed, including those carried out to employers' orders. The projects are presented to technical commissions [8].

Peter the Great St. Petersburg Polytechnic University (SPBPU). Implementation of modern approaches to training of engineers on the basis at the Institute for Advanced Manufacturing Technologies (substantive modernization of the mathematical foundation for engineers, practice-based learning technology in the framework of applied projects, creation of Digital Factory and others) [9].

Tomsk Polytechnic University. Preparation of the three major types of engineering specialists in TPU covers a wide range of training areas. The main components of the system for training engineering specialists (Figure 1) to meet the requirements of employers include:

– research and education centers and university laboratories (annual revenues from R&D is over 1.8 billion rubles);

– cooperation with enterprises (involvement of employers in the design of the basic educational program (BEP), study of specific modules focused on professionally-oriented practices, development of the project work themes, assessment of students' learning outcomes);

– increased amount of practices: 27 credits (18 weeks for 1-4 year);

– use of resources of other educational organizations (SSMU and others) to acquire unique competencies required in each field;

– implementation of the module "Preparation for an integrated engineering process" aimed to provide, the 1st and 2nd year students with a clear view of the future professional activity, to form the skills required for the project and team work, and to involve students in engineering practice through solving simple problems of design, modeling and analysis in the field of technology;

– 3rd and 4th year students deal with more complex projects to the order of enterprises within the educational and research work, and complex projects in professional disciplines aimed to prepare students for the graduate qualification work [8] (Figure 2).

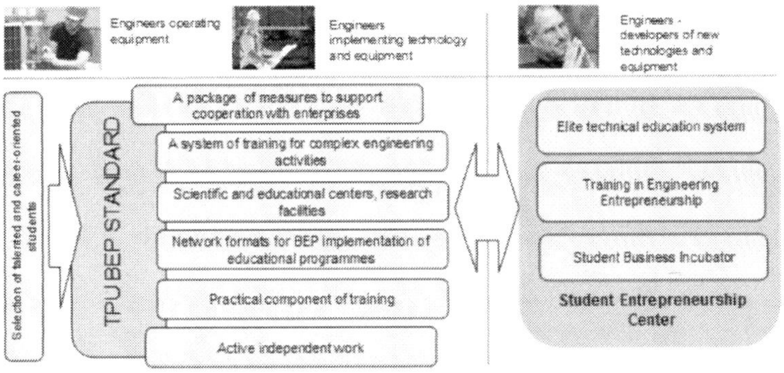

Fig. 1. Training of the personnel in TPU.

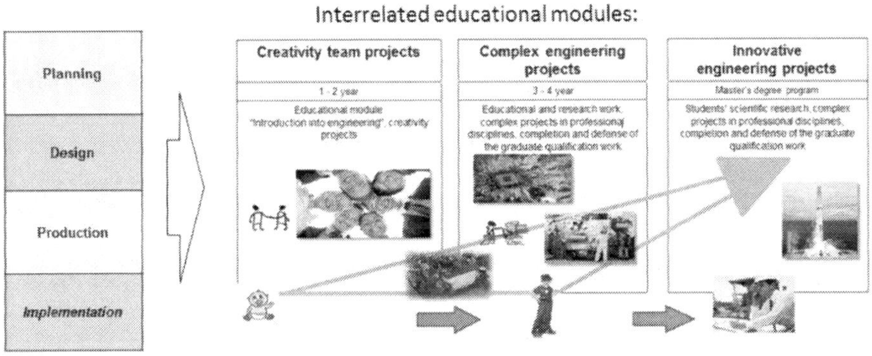

Fig. 2. System of training students for complex engineering activity.

To prepare specialists of the third type (capable of developing new technologies and equipment), an individual program has been implemented in TPU within the framework of elite engineering education since 2004 [10]. The program is developed for a group of selected students (about 10% of the total number of students). Similar educational programs are utilized in the world's leading universities: MIT, OLIN, University of Toronto and others. The elite engineering education program enables development of competences related to fundamental and professional knowledge, innovation, entrepreneurship and leadership in graduate students.

In addition, all students of undergraduate and specialty levels are enrolled in the discipline "Engineering entrepreneurship" (training in basics of entrepreneurship, acquiring skills in searching business opportunities and resources, new market niches, etc.). Moreover, competitions, trainings and business schools are held in the Student Business Incubator and the Department of Engineering Entrepreneurship.

4 Model of training specialists for high-tech industries

Based on the experience of Russian and foreign universities and needs of modern high-tech industries, a system of training specialists has been proposed to ensure the following activities (Figure 3):

- conduction of basic and applied research; active work with potential applicants of all levels of educational background (information events for career guidance and engineering promotion, involvement of talented students);
- students' project (including independent) work on real practice-oriented projects, a large proportion of practices within the scope of the program;
- implementation of a wide range of students' activities [11];
- entrepreneurship development system for students (special educational modules, trainings, resource support).

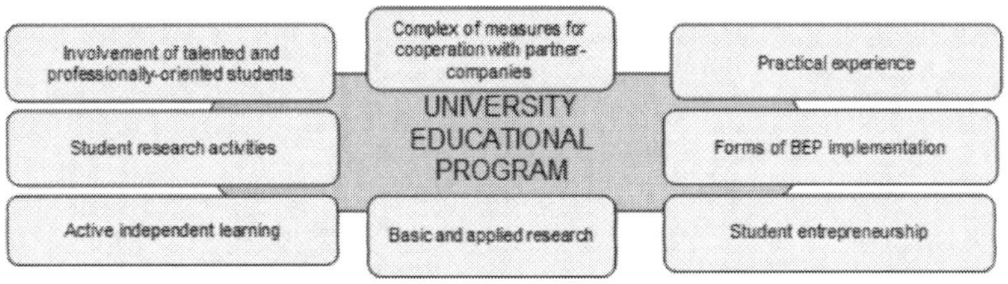

Fig. 3. Model of training specialists for high-tech industries.

The development of students' activity is crucial for acquisition of entrepreneurial competencies: independent learning, project work, students' entrepreneurship infrastructure (co-working centers, trainings, internal and external grants, acceleration programs, etc.). Implementation of the proposed principles and approaches will enable a modern university to create educational programs that are competitive in the global market of educational services [12].

5 Conclusion

The model of training sought-after specialists for high-tech industries (including those for the space industry) has been developed based on the strategy of science and technology development of the Russian Federation and the experience gained by Russian and foreign universities.

References

1. *Project 5-100*, URL: http://5top100.com/about/more-about/
2. *Strategy of scientific and technological development of the Russian* Federation, URL: http://sntr-rf.ru/
3. N. Daneikina, Y. Daneykin, *INTED2016 Proceedings*, 3565 (2016)
4. V. Ivanova, K. Mertins, et al, Mat. Web of Conf. 48, 06002 (2016)
5. *Speech of German Gref at the III Congress "Innovative practice: science plus business"*. URL: http://scienceplusbusiness.ru/congress (2016)
6. A.Ju. Filippovich, Ju.N. Filippovich, PREPOD. XXI VEK **3**, 28 (2016)
7. *National Technology Initiative*, URL: https://asi.ru/nti/

8. M. A. Soloviev, et al. Mat. Web of Conf. **48**, 06001 (2016)
9. A. Borovkov, *Digital Factor of Advanced Manufacturing Technologies Institute of SPBPU*. URL: http://fea.ru/news/6387 (2016)
10. O.M. Zamyatina, et al, WSJ 27 (ELELC), 434 (2013)
11. A. Shvindtl, I. Nikanorov, Mat. Web of Conf. (2017) (to be published)
12. G.A. Voronova, M.A. Soloviev, Molodoy Ucheniy, **10**, 589 (2015)

MATEC Web of Conferences **102**, 01035 (2017) DOI: 10.1051/matecconf/201710201035
V International Forum for Young Scientists "Space Engineering"

The technology of electromagnetic radiation danger estimation using the hardware-software module

Eugene Titov[1,*], and *Ivan* Migalyov[1]

[1]Polzunov Altai State Technical University, 656038 Barnaul, Russia

Abstract. The article describes the principles of functioning of the hardware-software module, whose purpose is the estimation of a danger level of the combined electromagnetic field influence on a human organism. The module consists of the hardware and the software parts. The hardware part is an array of electromagnetic parameter detectors; the software part is an electromagnetic field modelling program based on OpenEMS. The module creates so-called images of electromagnetic environment danger. The results show practical applicability of the technological module for the stated purpose.

1 Introduction

At present, the electromagnetic radiation created by technogenic electromagnetic radiation sources is a valuable factor of danger in both domestic and industrial applications [1-9].

The research conducted [6-8] shows that the measured electromagnetic radiation levels generated by operation of various electrical equipment may drastically exceed the maximal permissible level (MPL) of electromagnetic radiation on some frequencies. That means that simultaneous influence of multiple electromagnetic radiation sources should be taken into account when considering electromagnetic field parameters.

The research should solve the problem of estimating the level of the danger of the personnel staying in the zone of the influence of an electromagnetic radiation caused by multiple sources. The results will help to choose reasonably the measures to protect the personnel; these measures will be based on the new principles of multi-frequency control of the electromagnetic field parameters.

2 Experimental part

There is a special hardware-software module designed to automate the electromagnetic field measurement process and to create the images of danger in the zones of influence of multiple electromagnetic radiation sources on various frequencies. The module allows to monitor the measurement results in the real time, and to analyze the danger image of electromagnetic radiation.

* Corresponding author: 888tev888@mail.ru

© The Authors, published by EDP Sciences. This is an open access article distributed under the terms of the Creative Commons Attribution License 4.0 (http://creativecommons.org/licenses/by/4.0/).

The hardware-software module consists of the following blocks:

- a measurement devices block (one possible variant is ST-01, MTM-01, P3-50, BE-meter AT-004 and P3-41 devices to measure, correspondingly, electrostatic; magnetostatic; alternate electric and magnetic fields on industrial frequency, i.e. 50 Hz; electric and magnetic fields on radiofrequency, i.e. 30 kHz — 30 MHz; energy flow density on 300 MHz — 300 GHz, together with AKS-1201 spectrum analyzer);
- a block of device adapters designed to connect the measurement devices to PC;
- the specialized PC software to gather and analyze the measurement results.

Figure 1 shows the connection of measurement devices (basic configuration) to a PC.

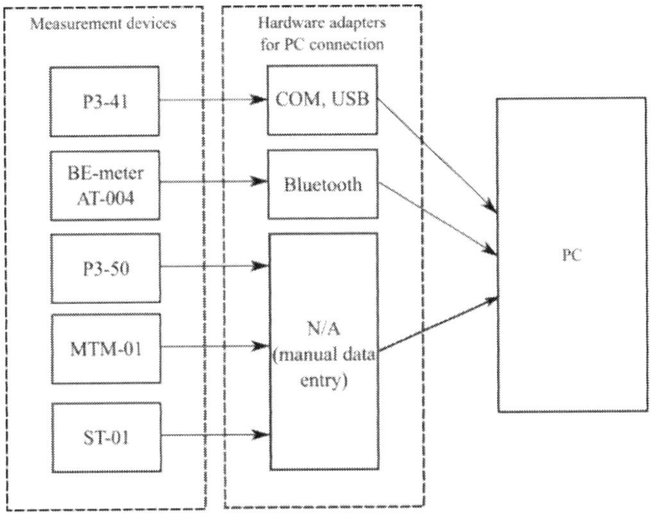

Fig. 1. Connection of measurement devices to PC.

PC saves the incoming measurement data into a separate database for every examination.

If the measurement device has the adapter that allows connecting it to the PC, it sends the experimental data directly through the adapter. Device manufacturers are not providing such adapters for some older devices, and, in that case, the hardware-software complex allows the user to enter the data manually from the PC keyboard. Existing data could also be converted to a format compatible with the complex.

After the measurement stage, the data is processed using the analytical software included into the hardware-software complex.

The functioning of the technological module is based on the following principles.

The digital model of the studied room (including all the electromagnetic radiation sources) is produced based on the geometric parameters of the facility, and the relative positions of the electromagnetic radiation sources. Every source is modeled as a 3D box; the model resolution should be 5 cm.

The measurement or every controlled electromagnetic field parameter (i.e. electrical and magnetic field values, energy flow density) is performed in every of the standardized frequency diapasons (i.e. 0 Hz, 50 Hz, 30 kHz — 300 GHz), including the subdiapasons (30 kHz – 3 MHz, 3 MHz – 30 MHz, 30 MHz – 50 MHz, 50 MHz – 300 MHz) and possibly higher frequencies. The measurement should be performed on the standard distance from every face of every electromagnetic radiation source in question; the standard distance should be determined by local sanitary rules and norms for every case. The main data collected on this stage is the maximal value of every measured parameter for every accessible face of every electromagnetic radiation source in the room.

For every frequency analyzed, prepare a computer model of the whole room, to derive the so-called electromagnetic field image for the whole room on this frequency.

AppCSXCAD program [10] used to create the 3D models allows to enter some of the electromagnetic parameters of the room on the room plan. Every electromagnetic radiation source and communication line in the room should be registered as a solid metal object.

3 Results and discussion

Each of the generated electromagnetic field spatial images is used to prepare the so-called electromagnetic field danger image. The hardware-software module achieves it by transforming the axe of the electromagnetic parameter (e.g. electric field, magnetic field, energy flow density) to the so-called allowed staying time (determined according to the local sanitary norms) axis in every image node.

The zones of the room where multiple danger zones are overlaying may be determined based on the analysis of the images of danger created by controlled components of electromagnetic field.

So-called cylindrical danger image is used to demonstrate the electromagnetic field danger inside of the rooms that have industrial purpose. The main difference between the point and cylindrical picture is the projection method used to prepare the picture. Every pixel of the cylindrical picture accounts the parameters of the electromagnetic field inside of the cylindrical zone (with some predetermined radius based on the industrial requirements) around the pixel. It helps to better consider the working zones of the personnel inside of the industrial room. The sample cylindrical picture of the room is presented in Figure 2.

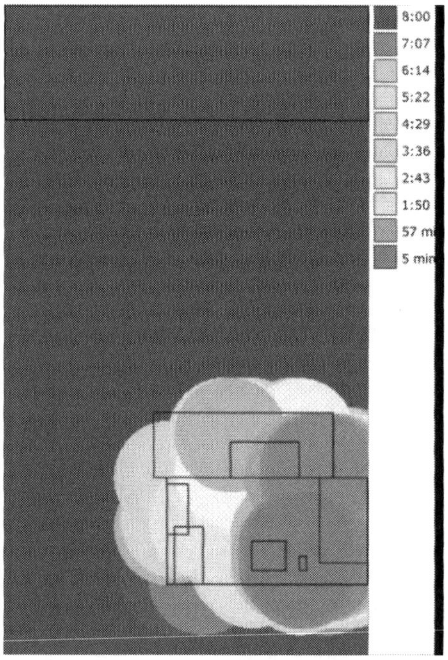

Fig. 2. Combined cylindric image of the electrolagnetic field (hours of allowed staying time).

4 Conclusion

The produced technological module allows to control the danger levels in the electromagnetic environments that include multiple electromagnetic radiation sources.

For the zones of the rooms with no overlays between the danger zones, the generated images are used to prepare the protective measures based on the values of the controlled electromagnetic field components in the controlled frequency diapasons.

For the zones with intersections of multiple frequencies, a complex danger combining algorithm should be used; one perspective model for that is the danger overlay model.

The resulting danger images with refined personnel stay time with the zones of complex electromagnetic radiation influence are used to derive the protection measures for personnel with respect to frequency diapason for every frequency in the studied room.

References

1. N. A. Volodina, R. N. Karyakin, L. V. Kulikova, O. K. Nikolsky, A. A. Soshnikov, *Basics of the electromagnetic compatibility* (2015)
2. I. I. Davydov, Computer technology, **1** (2004)
3. T. V. Veselova, V. A. Veselovsky, L. I. Deyev, A. A. Baizhumanov, Rad.Biol. Radioec. **52(4)** (2012)
4. M. A. Martínez, A. Úbeda, M. A. Cid, M. A. Trillo, Cell Physiol Biochem **29,** 5 (2012)
5. World Health Organization, *Electromagnetic fields and public health,* URL: http://www.who.int/mediacentre/factsheets/fs322/ru/ (2007)
6. V. N. Dovbysh, M. Yu. Maslov, Yu. M. Spodobayev, *Electromagnetic safety of electrical system elements: monography* (2009)
7. I. E. Migalyov, Polzunovskii Vestnik, **4** (2012)
8. E. V. Titov, *Increasing of agricultural complex's electrotechnology safety based on the integrated control of electromagnetic radiations: Ph. D. dissertation thesis* (2013)
9. R. G. Minullin, V. I. Nazarenko, E. Yu. Zykov, and others, Hygiene and sanitation **4** (1995)
10. T. Liebig, *AppCSXCAD – Minimal GUI Application using the QCSXCAD library,* URL: https://github.com/thliebig/AppCSXCAD

Measuring coordinates of objects with adaptation expansion options

Aleksandr Tsitsulin[1,*], *Viacheslav* Piatkov[1], *Gennadiy* Levko[1], and *Aleksey* Morozov[1]

[1]Joint-stock company «Institute of Television», Scientific-Technical Complex, 194021Saint-Petersburg, Russia

Abstract. The parametric optimization of coordinates measurement with variations of definition and frame rate by limiting the readout speed of the signal from the image sensor problem is considered. It is shown that the optimal values of definition and frame rate are determined by the alignment of the inter-element and inter-frame difference variances estimates; switching thresholds can be decreased by the delay introducing into the mechanism of measuring these dispersions to increase the observation time with close values of the estimates of these variances.

1 Introduction

More than half a century spacecrafts (SC) have been docking in orbit with usage of radio and optical-electronic measuring systems (OES) intended to measure the time-varying spatial coordinates of cooperating spacecrafts. To ensure the implementation of the SC convergence method it is necessary to measure the coordinates of the guidance object. An important principle of the coordinate measuring of that object, especially at long range is the principle of the tracking measurement. An important feature of SC convergence and docking control that was introduced by S.P. Korolev is principle iterative coordinate measurement: changing distance between SC leads to changing parameters of measuring system. In last few years, this principle is complemented by the method of adaptation of image decomposition parameters using solid-state matrix image sensors [1-3]. This trend in the construction of measuring systems is connected on the one hand with the change in signal statistics over time (due to changes in distance between the SC), on the other hand - the restriction of signal readout speed from matrix image sensors. Task of finding spatiotemporal coordinates total estimation error minimum refers to isoperimetric problems, where partial errors estimation $\varepsilon_k = D_{\Delta k}$ act as variables, and available reading speed of sensor – as a «perimeter». To solve this type of problem Euler equation with undefined Lagrange multipliers is used. The solution of the equation is similar to the solution of error distribution in the signal spectrum problem (the calculation of the Epsilon-entropy) and leads [1-3] to the concept: for each plot (target-background environment), there is a best video from which is derived the rule of variances equality increment of signal by all the arguments (the rule of "equilibrium" increments). In the simplest case of the image

* Corresponding author: niitv@niitv.ru

© The Authors, published by EDP Sciences. This is an open access article distributed under the terms of the Creative Commons Attribution License 4.0 (http://creativecommons.org/licenses/by/4.0/).

observation in a single spectral range this rule corresponds to the requirement of variance equality inter-element (horizontally and vertically) and a frame difference of videosignal: $D_{\Delta x} = D_{\Delta y} = D_{\Delta t}$. Such equation, that supposing measurement procedure of increments dispersions for different coordinates allows minimizing the miss in spatiotemporal coordinates of SC.

2 Discussion

The presence of such a conditional decomposition parameters optimum intuitively clear, because apparent size and speed of the object are small at the long-range, and the ultimate clarity is actual and in the low range size and speed of the object are high, and more relevant parameter is high frame rate.

Thus, the system of coordinates measurement, that implements the alignment by means of definition (number of elements of decomposition) and frame rate *interchange*, includes the measurement subsystem of z statistics (relative to the current ratio of the inter-element and inter-frame variances estimates $z = D_{\Delta x}/D_{\Delta t}$). It is necessary to use traditional methods to ensure the stability of the automatic control system as well as new methods based on non-linearity and persistence in the third closed loop [1-3]. Due to the discreteness of the matrix image sensors a change in image sharpness can only be done discretely which makes its specificity in the synthesis of control systems. When measuring coordinates of the object in each *k*-frame an alternative of TV image quality arises. In the simplest case, the adaptive system to the dynamic plot can have two states:

- $z > \gamma_{high}$ transition to a state with maximum clarity;
- $z < \gamma_{low}$ transition to a state with maximum frame rate;
- $\gamma_{low} \leq z \leq \gamma_{high} \rightarrow$ preservation of the existing state.

The minimum ratio threshold $d = \gamma_{high}/\gamma_{low}$ corresponds to a change intervals relations of spatial and temporal sampling, which is equal to 4 in this case. However, using thresholds $d = \gamma_{high}/\gamma_{low} = 4$ for stability of the adaptive system is not enough, due to the discreteness of the raster, and inevitable presence of noise in the video signal. Measurements show that the monotonous variation of range in this problem gives monotonous variation of *z*-statistics. At the same time in the measurement of fluctuations relative to the expectation of the smoothed function $f(t)$ [3] are identified, that determines the necessary of incrimination of the thresholds ratio $d = \gamma_{high}/\gamma_{low}$, which ensures the stability of automatic control system. If we assume that the observed signal can be modeled by a stochastic process with a normal distribution, including uncorrelated signal and noise components, the *z* statistics becomes a modified Fischer-Snedecor statistics [2]. This distribution $p(z|f)$ for a fixed mathematical expectation f of *z*-statistics for large number of samples (corresponding to the number of elements in the image sensor array, which is $10^5 \ldots 10^7$ typically) is symmetrical. The calculation results show that the analysis in real time of each frame (without delay) for reasonable values of the probability of false alarm (untimely changes in the system state is fraught with immediate reverse change; $p_{fa} \leq 10^{-3}$) thresholds differ significantly from the minimum, which can be obtained from image discreteness and thresholds ratio $d_{1\leftrightarrow 2} = \gamma_{high}/\gamma_{low}$, in the absence of fluctuations is equal to 4, must be increased approximately to 9 [2].

The high function quality of the objects coordinate measuring OES is provided by decomposition parameters adaptation based on the maintenance (approximate, with a certain confidence interval) of video-signal increments dispersions equality by all of the arguments.

Next, the simplest case of a system with two states is considered. In real systems, expectation rate of change $f(t)$ of reference *z*-statistics is small, and the deviation of *z* statistics from the mean value $f(t)$ are statistically independent in the adjacent frames, whereby it becomes possible to minimize the time when the system is in the "wrong" state (when different arguments signal increment variances differ), or for the simplest embodiment of a system with two states – the possibility of increasing the residence time of the system in a state of $\frac{1}{2} < z < 2$. This problem relates to the optimal filtering problems (non-linear generally), in which the evaluation of the current value of the derivative of the mathematical expectation $f(t)$ of z-statistics in time $df(t)/dt$ are formed. Function $f(t)$ can be modeled by a piecewise-linear relationship (in the real docking conditions correcting engine are working, which introduce additive jumps in the variation of $f(t)$). Considering that $f(t)$ is a monotonous function, it is possible to obtain an assessment using the optimal Calman filter, introducing a variable time inertia in the estimation process. This version of the problem of obtaining the optimal inertia control system methodologically akin to obtaining the linear filter's optimal inertia in the Wiener-Kolmogorov's filtering theory. This is evident from the fact that, as in the linear filtering, there are two opposite effects: the introduction of inertia reduces the noise error, but increases latency (the analog of the linear error).

Then, considering the discrete form of the measurements formation $f(n)$, where $n = 0, 1, 2 \ldots$ (n corresponds to the moments of time t_n), state vector $\mathbf{f}(n) = \left[f(n) \quad \dot{f}(n)\right]^{\mathrm{T}}$ the generation algorithm of estimates vector parameters $\hat{\mathbf{f}}(n)$, is of the form:

$$\hat{\mathbf{f}}(n) = \mathbf{\Phi}(n)\hat{\mathbf{f}}(n-1) + \mathbf{K}(n)\Delta f(n), \quad \hat{\mathbf{f}}(0) = \hat{\mathbf{f}}_0, \qquad (1)$$

$$\Delta f(n) = f(n) - \mathbf{C}\Phi(n)\hat{\mathbf{f}}(n-1).$$

where $\mathbf{\Phi}(n)$ – 2×1 extrapolation matrix; $\mathbf{K}(n)$ weight matrix; $\Delta f(n)$ – fault; $\mathbf{C} = [1\ 0]$ – 1×2 observation row matrix, showing that only one value $f(t)$ is measured.

Based on analysis and equations for Kalman's filter weight matrix K(n) transformation the new equations can be obtained:

$$\left.\begin{aligned} \mathbf{K}(n) &= \mathbf{P}_e(n)\mathbf{C}^{\mathrm{T}}[\mathbf{C}\mathbf{P}_e(n)\mathbf{C}^{\mathrm{T}} + R(n)]^{-1}, \ \mathbf{P}_e(1) = \mathbf{P}_1; \\ \mathbf{P}(n) &= \mathbf{P}_e(n) - \underline{\mathbf{K}(n)\mathbf{C}\mathbf{P}_e(n)}; \\ \mathbf{P}_e(n+1) &= \mathbf{K}(n)\mathbf{K}^{\mathrm{T}}(n)\overline{(\Delta f(n))^2} + \mathbf{P}(n). \end{aligned}\right\} \qquad (2)$$

Here $\mathbf{P}_e(n)$ and $\mathbf{P}(n)$ – 2×2 extrapolation and estimation errors covariance matrixes respectively; $R(n)$ – measurement error covariance matrix, which is scalar at indicated above observation matrix $\mathbf{C}$; «T» – transposition sign. The value $\overline{(\Delta f(n))^2}$ should be determined statistically over several frames.

Estimation algorithm for parameters of vector $\mathbf{f}(n)$ by the expressions (1) и (2) defines the structure and parameters of the third control loop (image sensors' parameters of decomposition) and allows the system to minimize the statistic's $f(t)$ estimation errors on it's linear sections, as well as on "jump like" variations of $df(t)/dt$, responding to any changes in the input value $f(t)$ by changing component values of the weight matrix $\mathbf{K}(n)$.

Simplified version of this filtration can be implemented using a nonlinear accumulation, i.e. making a decision after reference z-statistic exceeds reduced threshold in several frames. In this simple case there is optimum amount m jointly analyzed serial frames, depending on current value of the derivative of the mathematical expectation f of z-statistics in time $df(t)/dt$. Therefore, to minimize system "wrong" state spent time an extremum of the functional should be found:

$$t_m = m + \frac{\gamma_m - \gamma_\infty}{\operatorname{arctg}\left(df(t)/dt\right)} \to \min. \quad (3)$$

Analytic solution of the problem (3) is possible when threshold dependence $\gamma_m(m)$ is clearly formalized. This dependence can't be found in analytical functions, and for the required value of optimal number of frames evaluation the function approximation $\gamma_m(m)$ can be used. Numerical analysis has shown that a good approximation of this dependence is a function $\gamma_m(m) \approx \gamma_\infty + m^{-1/4}$. Then, after taking the derivative dt_m/dm and equating it to zero, we can find the optimal number of frames m, for which solutions about z-statistics threshold γ_m excision should be multiplied:

$$m_0 \approx \left[4\operatorname{arctg}(df(t)/dt)\right]^{-4/5}. \quad (4)$$

Formula (4) calculations (small values $df(t)/dt$ simplify the formula) shows that when $df(t)/dt > 10^{-1}$ optimal inertia is only 2 frames, and for average observation distance at typical SC docking problem [3] $df(t)/dt = 10^{-2}...10^{-3}$ inertia is several tens of frames. For these values borders of threshold ratio $d_{1\leftrightarrow 2}$ of making a decision to change imaging mode are about 5.8…5.5, which is not 4 or 9, as we get in absence of inertia.

3 Conclusions

Thus, the solution to the problem of minimizing measurement errors for varying in time objects coordinates with limited readout speed from image sensors allows finding conventional optimum definition and framing rate for discretely switchable interchange. Achieving sustainability of the automatic control system by the measurement system's decomposition parameters requires the introduction of confidence interval when measuring statistics of the signal that exceeds minimum allowed value determined by the discrete image.

References

1. A.A. Umbitaliev, A. K. Tsitsulin, A. A. Mantsvetov and etc., J OPT TECHNOL **11**, 84 (2012)
2. A.A. Umbitaliev, V. V. Piatkov, A. K. Tsitsulin and etc., J Ques RE TV EQUIP **3**, (2015)
3. A.V. Morozov, V. V. Piatkov, G.V. Levko and etc., J Ques RE TV EQUIP **1**, 3 (2016)

Multiphase resonant inverter with sine wave output voltage

Irina Vavilova[1,*], and *Dmitriy* Ogorodnikov[1]

[1]National Research Tomsk Polytechnic University, 634050 Tomsk, Russia

Abstract. Methods to obtain the sinusoidal output voltage in DC-AC converters are considered. Large switching losses in power switches, high frequency interference, and the relatively high cost are the major disadvantages of DC-AC converters with pulse-width modulation. The circuit of a multiphase parallel resonant inverter with a pure sine wave output voltage is proposed. The operating principle of the two-phase inverter and the method of output voltage control are described. The results of resonant inverters OrCAD simulation are shown. The advantages of the proposed inverter circuit are shown in comparison with a single-phase circuit.

1 Introduction

The main part of the electronic equipment can be AC powered. But in cases where industrial network is not available, there is a necessity for devices that convert the primary power supply voltage to the AC network voltage. In most cases, as it is known, the primary sources do not satisfy the consumer's with frequency and magnitude of the output voltage or with the stability of its main parameters. Most often relatively low DC voltage of the primary source should be converted to a relatively high AC voltage. A power inverter, or an inverter, is an electronic device that changes the direct current to the alternating current. Since the inverter power switching devices (transistors, thyristors) generally operate in the switching mode, the natural form of the output voltage is a rectangular one. However, to achieve the flexibility of the inverter with respect to the AC loads, the inverter must generate the sine wave output voltage. Nowadays the idea of renewable energy is very popular, particularly solar and wind energy, so there is a constant need for devices which convert the unstabilized relative low DC voltage from wind turbines or solar panels to a relatively high sine wave voltage.

2 Materials and methods

The purpose of this study is to design a model of a multiphase circuit that controls the output sinusoidal voltage of the resonant inverter.

DC to AC conversion technologies received a significant development in recent years.

* Corresponding author: irenavavilova@gmail.com

© The Authors, published by EDP Sciences. This is an open access article distributed under the terms of the Creative Commons Attribution License 4.0 (http://creativecommons.org/licenses/by/4.0/).

Modern inverters provide a sinusoidal output voltage with low harmonic distortion, and low levels of electromagnetic interference [1]. Very often the inverter is used as one of elements in the power converter. The principal feature of such schemes is the increased operating frequency (up to several hundreds of kilohertz). For efficient operation of the converter at a high frequency more sophisticated and expensive electronic components are required.

The power converter must have a high efficiency, high reliability and have acceptable weight and size. It must also have a valid level of the higher harmonics (THD limits) in the output voltage and do not create unacceptable to the consumer level of output voltage ripples. Voltage inverters help information systems to eliminate or to reduce the dependence on operation of the AC main. For example, when electric power of the main goes off unexpectedly the backup battery paired with the inverter will ensure the operation of computers until the execution of the necessary tasks will be completed correctly.

Typically, the pulse width modulation (PWM) is used to control the inverter output voltage and the output frequency. The output voltage in such schemes has a pulse shape. However, a significant switching loss is the main disadvantage of the PWM inverters. One way to reduce the switching losses is to design the inverter circuit so that the switching is done at zero current or zero voltage (so-called "soft switching" technique). For example, sinusoidal nature of current in the resonant inverter provides a number of advantages: minimal switching losses of power switches, high reliability, low levels of radio-frequency interference (RFI), the possibility to use thyristors – devices for which only the turning on is controlled.

Resonant circuits are used in high-power thyristor motor drives and uninterruptible power supplies. However, they are rarely used in low-power converters.

In cases when the load varies over a wide range (including open circuit case), the preference is given to the parallel resonant inverter. Besides the advantages mentioned above, the work in the series resonant circuit at resonance mode also allows receiving an increased output in comparison with the input voltage without using a transformer.

Despite the advantages, only simple circuits of resonant inverters are discussed in the literature. These schemes are not widely used as the sine wave generators. In particular, this happened due to the rather strong dependence of the output voltage on the oscillating circuit quality factor.

2.1 Resonant inverters

The circuit of the parallel resonant inverter providing power consumption regulation and, as a consequence, the stabilization of the output voltage is shown in Figure 1 [2, 3].

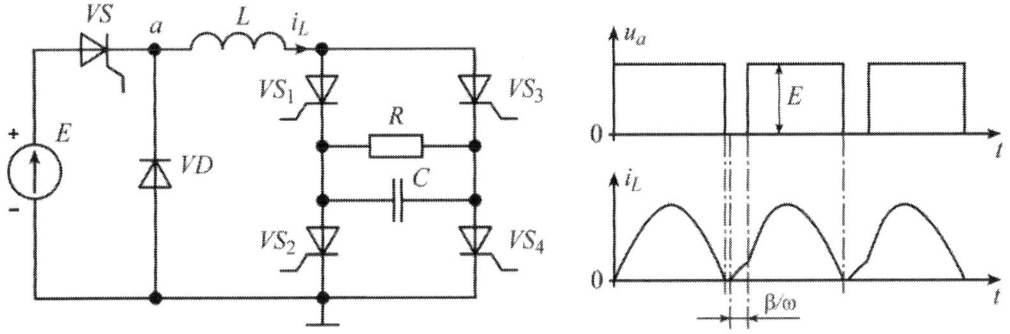

Fig. 1. The parallel resonant inverter with the input thyristor *VS* and its voltage and current-waveforms.

The pulsed input source energy consumption in the process of output voltage regulation requires the use of a smoothing filter. To reduce the weight and size parameters of this filter we need to reduce the form factor of the consumed current. The main idea is that in order to reduce the form factor of the current is expedient to consume the power from the input source through several parallel phases. These phases are included in the operation as needed (the switching time for the next phase depends on the load of the inverter). With this consumption method, the input current amplitude will be smaller and, hence, longer duration than in the single-channel circuit.

2.2 Multiphase circuit

The developed circuit to realize the proposed energy consumption principle is shown in Figure 2. In this example, the input circuit is divided into two phases (such circuits are called multiphase).

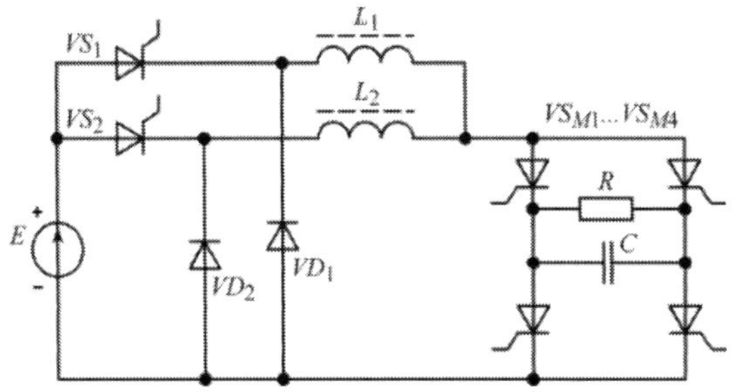

Fig. 2. The biphase circuit of a parallel resonant inverter.

In the first cycle of the inverter operation both input thyristors VS_1, VS_2 and a pair of thyristors in one of the diagonals of the H-bridge are switched on. The charge of the capacitor is influenced by the voltage source E. The difference between the input voltage and the voltage across the bridge diagonal is applied to the input inductors, and they are connected in parallel to each other. Hence it follows that to obtain the desired output frequency and Q-factor of this circuit the equivalent inductance of L_1 and L_2 inductors which are connected in parallel should be equal to the inductance L of the original single-phase circuit. Consequently, the inductance of each inductor should be two times higher than the inductance of a single-phase circuit. If L_1 and L_2 inductance are equal, the amplitude of the current in each phase will be equal to half of the current amplitude in the single-phase circuit, and the phase amplitudes are equal to each other.

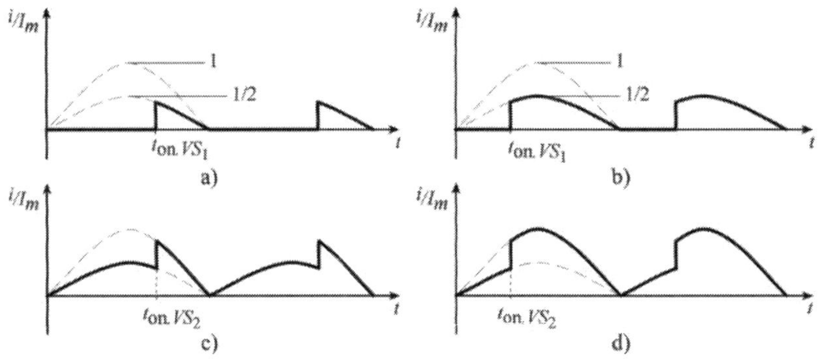

Fig. 3. The power supply current waveform with increase of the load of the biphase inverter circuit, where I_m is the amplitude of the power supply current at the maximum load.

The switch on time delay (firing angle) of the input thyristor with respect to the switch on time of H-bridge thyristors will be the smaller, the higher the inverter load is. At the light load mode the regulation (or stabilization) of the output voltage is provided only by one phase. But if the thyristor VS_1 already conducts during the whole period, and the power in the load is still not compensated, then the second phase (VS_2) starts to operate. Then thyristor VS_1 switches on simultaneously with the H-bridge thyristors (firing angle is zero), and the second phase thyristor VS_2 switches on with a delay. Thus, firing angles of the thyristors VS_1, VS_2 vary from 0 to π depending on the load, wherein the operation of the second phase thyristor takes place after the zero firing angle of the first phase input thyristor. The Figure 3 shows how the power supply current waveform changes with increasing the inverter load: from the light load (Figure 3, a) to the almost maximum load (Figure 3, d).

3 Results and Discussion

The study of a model of the biphase resonant inverter circuit using OrCAD, confirmed the correctness of our assumptions. PID controller is used to stabilize the output voltage. The output voltage and the power supply current waveforms are shown in Figure 4.

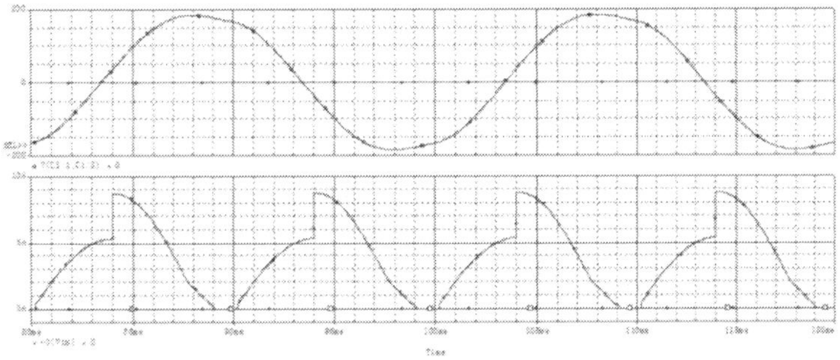

Fig. 4. The output voltage and the power supply current waveforms (steady-state operating mode, output voltage THD is about 5%).

The research shows that in a large range of firing angles (which corresponded to load change from about 3% to 80% of maximum) the advantage in RMS consumed current is not less than 20% compared with single phase resonant inverter. The maximum reduction of the form factor of the consumed current is about 67%. Transistors as fully controlled switches can be used instead of thyristors as input switches.

4 Conclusions

The biphase circuit provides not only decrease of the form factor, but also a decrease of the amplitudes of higher harmonics in the consumed current, which also would reduce the weight and size of the smoothing filter of the inverter.

References

1. V.M. Bushuev, V.A. Demyanskij, L.F. Zaharov i dr. *Elektropitanie ustrojstv i sistem telekommunikacij: Uchebnoe posobie dlya vuzov* (M.: Goryachaya liniya – Telekom, 2009) (in Russian)
2. D.N. Ogorodnikov, E.V. Yaroslavtsev, Instrum. Experim. Tech., **42(3)**, 383 (1999)
3. D.N. Ogorodnikov, *6th Annual International Siberian Workshop and Tutorials on Electron Devices and Materials (EDM)*, 1523212 (2005)

Determination of the weld thickness of turbine for aircraft engine by high-energy X-ray tomography

Yang Zhong[1], *Sergei* Chakhlov[1,*], and *Talgat* Mamyrbayev[1]

[1]Tomsk Polytechnic University, 634050 Tomsk, Russia

Abstract. It is necessary to test the weld thickness of turbine, as it is one of the most important parts for aircraft engine. The weld thickness of turbine for aircraft engine by high-energy X-ray tomography was determined. We used an X-ray tube and a betatron as X-ray sources. The wall thickness of two tubes and weld thickness of turbine were measured. It is shown that the high-energy X-ray tomography system is determined the wall thickness of the tube and the weld thickness of turbine with high accuracy. We also studied the method to reduce scattered radiation. All experiments were carried out in the non-destructive testing (NDT) Institute of Tomsk Polytechnic University (TPU).

1 Introduction

Recently the welds inspection is very popular subject of many researches [1-2]. To ensure the reliability of turbine for aircraft engine, it is necessary to test the weld thickness. In particular, the determination of the weld thickness is required to localize defects in the weld.

The comparative work [3] shows that the ultrasonic testing is not appropriate to identify typical micron-sized friction defects for linear welding. The only acceptable non-destructive method to control the internal defects is computed tomography (CT). The work [4] shows that by evaluating the state of defectiveness, we can define the period of the turbine operating. Many of similar studies [5-7] were conducted to determine the quality of welding in a turbine blade.

The purpose of non-destructive testing of the weld thickness of turbine is not only a determination of weld thickness and defects, but also recognition of their way to assess the potential danger of the defect.

2 Materials and methods

To determine the weld thickness of turbine for aircraft engine, we use high-energy X-ray tomography. The system of high-energy X-ray CT consists of a radiation source (X-ray tube or betatron), movement system, detector (line array), control system and image

[*] Corresponding author: chakhlov@tpu.ru

© The Authors, published by EDP Sciences. This is an open access article distributed under the terms of the Creative Commons Attribution License 4.0 (http://creativecommons.org/licenses/by/4.0/).

reconstruction system. The high-energy X-ray CT system was established in NDT Institute of TPU [8].

In our case, the object width exceeds the detector width, regarding of this, we use partial scanning method [9] for a complete image. In view of [10], the principle of scanning is that the symmetry axis of the object does not coincide with the rotation axes.

In order to verify the measurement accuracy, we measured the wall thickness of two tubes at first. Then we measured the weld thickness of the turbine. Tubes and turbine are shown in Figure 1.

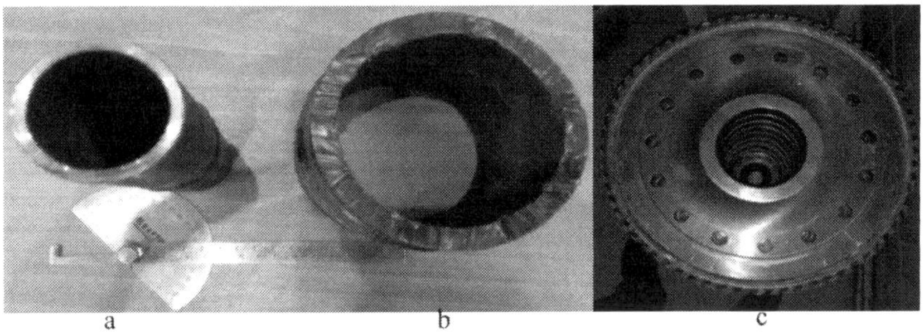

a b c

Fig. 1. a – Tube 1: outer diameter 47 mm, wall thickness 4 mm. b - Tube 2: outer diameter 115 mm, wall thickness 11 mm. c - Turbine: outer diameter 600 mm, wall thickness 20-30mm.

3 Results and discussion

The parameters of scanning modes for tubes and turbine are given in Table 1.

Table 1. Parameters of scanning modes for tubes and turbine.

Object	Accelerating voltage / kV	Tube current / mA	Copper filter thickness / mm	Number of projections	Angular pitch /degree
Tube1	300	1600	2	4800	0.051
Tube2	450	1700	2	4800	0.051
Turbine	450	1700	4	4800	0.051

The clear external boundary of the tube we can see in Figure 2. The internal boundary is more noticeable in Figure 2d in compare with Figure 2a. It was resulted due to the increase of tube current and corresponds to the work [11]. After segmentation, we get the inner and outer radius of the tubes, the wall thickness and relative error, presented in Table 2.

Table 2. Measured sizes of tubes and their errors.

Object	Inner radius / mm	Outer radius / mm	Wall thickness / mm	Standard value / mm	Relative error / %
Tube1	19.32	23.49	4.17	4	4.25
Tube2	48.36	59.69	11.33	11	3.0

Figures 2 and 3 show images obtained by use VGStudio MAX 2.2.

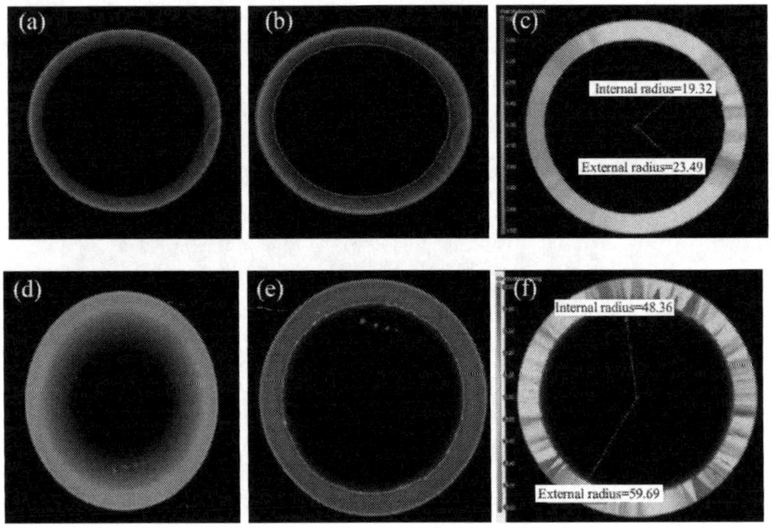

Fig. 2. a,d - 2D tomographic slice; b,e – Segmentation; c,f - Measurement of internal and external radius.

From the radiography Figure 3a, it was selected one sector of welding and shown in Figure 3b. The boundaries and defects shown in Figure 3c.

In order to reduce scattered radiation and obtain a clear inner boundary of the tube, we use gravel as absorber. Scanning mode was the same as in Table 1 for turbine. The result image is shown in Figure 4. In comparison with Figure 2d, the inner boundary became more pronounced.

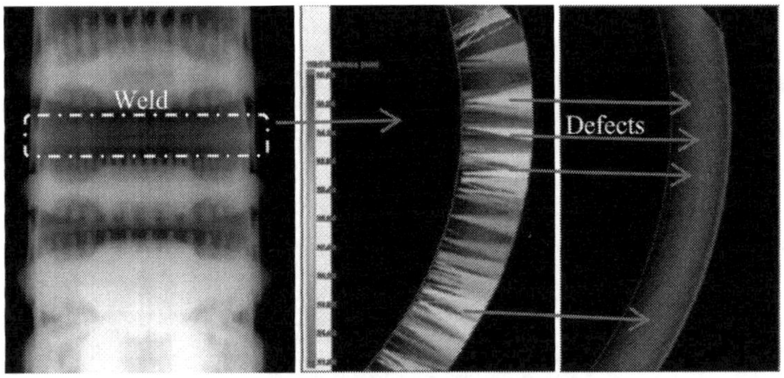

Fig. 3. a - Radiography, b - The definition of the weld thickness, c – Defects in weld joint.

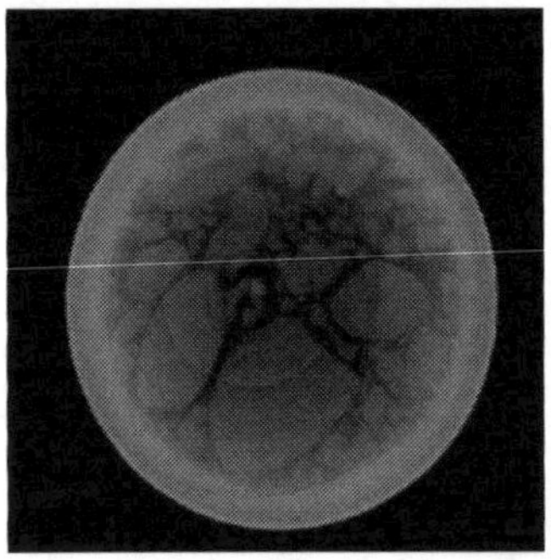

Fig. 4. Tomographic slice with gravel.

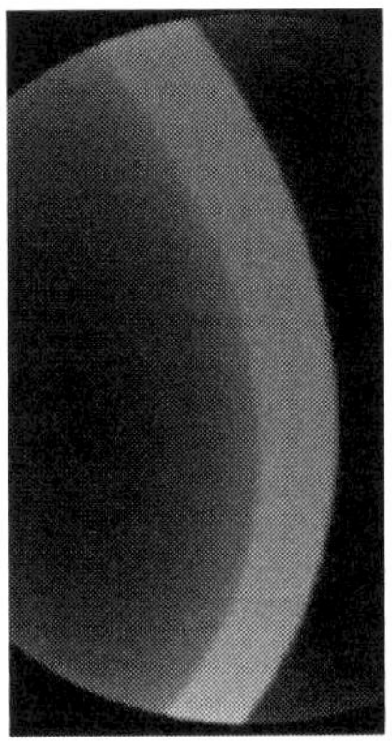

Fig. 5. A slice of turbine sector produced by the betatron tomography.

Figure 5 shows one sector of turbine, reconstructed for energy of betatron 3.5 MeV and for angle pitch 0.18 degrees. One can see that the inner boundary of the turbine is more clearly in comparison with Figure 3c.

4 Conclusions

Experimental results indicate that the high-energy X-ray CT system can determine the wall thickness of the tube and the weld thickness of turbine with high accuracy. The weld joint defects are identified by the tomographic measurement of the weld thickness. That can help to enhance the turbine welding technology and to prolong the turbine service.

We get a clear inner boundary of the tube by using gravel as absorber or betatron as radiation source. These results indicate that using absorber or increase the scanning energy are feasible methods to improve the image quality.

References

1. V.N. Borikov, O.V. Galtseva, G.A. Filippov, Journal of Physics: Conference Series **671(1)**, 012014, (2016)
2. S.V. Muravyov, V.N. Borikov, S.A. Kaysanov, *18th IMEKO World Congress* **1**, 77 (2006)
3. V.M. Bychkov, A.U. Medvedev, M.P. Savich, V.V. Smirnov, A.V. Supov, Engineering, **7** (2012)
4. L.K. Generals, M.I. Mochalov, Series: Natural and Tech. Sciences **6** (2015)
5. Zh. Chaozong, G. Zhiping, Z. Peng. *Industrial CT techniques and principles* (Science Press, 2009)
6. K.M. Oluwasegun, C. Cooper, Y.L. Chiu, I.P. Jones, H.Y. Li, N.G. Baxter, Mater. Science. & Eng. A. **596** (2014)
7. J. Aschenbrucka, R. Adamczuka, J.R. Seumea, 3rd Intern. Conf. Through-life Engineering Services Session: Recent Progress in Jet-Engine Regeneration (2014)
8. High-energy X-ray CT system. URL: http://portal.tpu.ru/departments/laboratory/rknl/products/big_tomo (2017)
9. Standard 7512-82. *Nondestructive testing. Welded joints. Radiography method*
10. A.V. Nissan, Technology in the electronics industry, **3** (2011)
11. D. Fleischmann, F.E. Boas, Imaging Med, **4** (2012)

MATEC Web of Conferences **102**, 01039 (2017) DOI: 10.1051/matecconf/201710201039
V International Forum for Young Scientists "Space Engineering"

The increase of flexible protective materials strength characteristics by electrophysical influences

Irina Zlobina[1,*], and *Nikolai* Bekrenev[1]

[1]Yuri Gagarin Saratov State Technical University, 410054 Saratov, Russia

Abstract. We studied the strength of the individual threads in the longitudinal and transverse rupture, the state of threads, as well as the strength of a single layer of aramid fabric exposed to the piercing influence of the conical indenter. Materials used in the experiments: special high molecular material of the brand DZh (SHMM-DZh) and ultra-high molecular weight polyethylene (UHMWP). It is found that the effect of electromagnetic field on the aramid threads and fabrics based on them, as well as the threads made of UMWP increases the tensile strength at 2 and 1.6-1.7 times respectively. Aramid fabric puncture resistance in the delivery state and in the wet state, after exposure to the electromagnetic field increases by 1.7-1.8 times. A significant reduction of the fiber bundle processed in the electromagnetic field of aramid fibers and UHMWP after exposure to tensile load was ascertained.

1 Introduction

The analysis of scientific literature, materials of conferences and exhibitions testifies the intensive development of the production of composite materials based on carbon fibers and glass fabrics [1, 2]. Aramid threads in comparison with carbon and glass fibers are less fragile and suitable for processing on conventional equipment of textile manufactures [3, 4].

Research in the field of increasing the strength of such fabrics is aimed at improving technologies for the production of the starting components and the formation of tissue by optimizing the weaving [5], which makes the process difficult and expensive, and also leads to the need for adjustment or re-production, which lengthens the cycle of introducing the new developments.

Other promising material, used in protective equipment, both independently and as internal support for ceramic armor plates - UMWP, formed from woven fibers [6-8].

[*] Corresponding author: irinka_7_@mail.ru

© The Authors, published by EDP Sciences. This is an open access article distributed under the terms of the Creative Commons Attribution License 4.0 (http://creativecommons.org/licenses/by/4.0/).

Previously, we have obtained positive results in increasing the strength of structural composite materials based on a polymer matrix reinforced with carbon fibers, after electrophysical effects, in particular, an electromagnetic field of a certain frequency and intensity [9, 10].

The purpose of the research is to study the possibility of improving the performance properties, namely, strength of aramid and UHMWP materials when exposed to an electromagnetic field to improve their safety performance in extreme conditions, adjoint with a mechanical effect on the human.

2 Materials and methods

Experiments were carried out on fabrics and threads (SHMM-DZh) and UHMWPE threads. Tissue samples TSVM-G were tested for puncture in dry and wet condition. Threads of both types of materials were tested for rupture. In all the experiments on investigation of the physical and mechanical properties one-layer SHMM-DZh tissue samples of 20 x 30 mm and threads 0.05 mm thick and 70 mm long were used.

Special equipment developed by us enables testing of samples puncture and tear. In the experiments a concentrated stream of low energy, medium and high intensity was generated. In all experiments, the processing time was 1, 4 and 8 minutes. 3 samples were processed at each intensity level. With a cone angle of the 900 loading device we influenced the tissue measuring the current force is taken from the load cell. The puncture moment was fixed by the load torque chart`s drawdown. The tests were conducted for dry and wet samples.

During electrophysical processing of the SHMM-DZh thread medium intensity mode was used, and low intensity mode for the UHMWP thread. The processing time took 4 minutes.

3 Results and discussion

Graphs showing the pressure increase kinetics until the moment of dry and wet samples puncture processed at an average intensity of electrophysical effects for 4 minutes are shown in Figure 1.

The research of the samples appearance after the puncture has showed that after electrophysical treatment both dry and wet samples puncture field form had a pronounced conical imprint, while in the control wet sample the loss of form was clearly noticeable, indicating the sliding and displacement of fibers, i.e. violation of their original weaving.

Analysis of the graphs (Figure 1) led to the conclusion that both wet control and processed samples have reduced at 1.5 times the puncture pressure compared with the dry sample. However, the wet sample, previously processed has 1.7 - 1.8 times higher pressure puncture strength compared to the control one. Moreover, its strength is even 1.25 times higher than the control sample in the delivery condition. Tests of SHMM-DZh and UHMWP threads showed the following. Load increase kinetics of the sample with load applying to the SHMM-DZh thread in the axial direction is shown in Figure 2, and the UHMWP threads - in Figure 3.

The graphs show the different nature of the destruction of threads made of the studied materials. Aramid threads tear almost immediately, the graphics are of a smooth shape. UHMWP threads tear` charts have serrate character and show periodic load`s bursts and falls. Analysis of the graphs shows that electrophysical processing at the average intensity of the impact increases the axial tensile load on the SHMM-DZh thread almost in 2 times. The influence of electrophysical processing on the thread`s strength is significantly less pronounced in the cross direction. The tensile strength in the transverse direction does not increase by more than 18-20%.

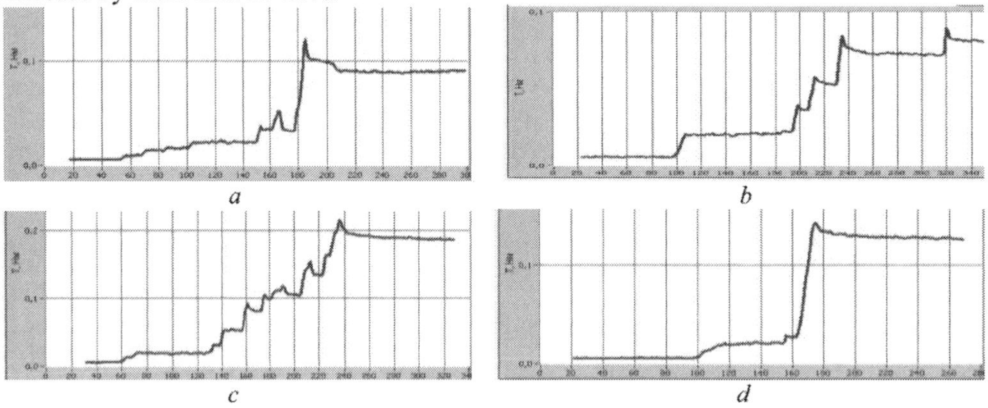

Fig. 1. Samples puncture effort` kinetics: control dry (a) and wet (b); dry processed (c) and wet (d).

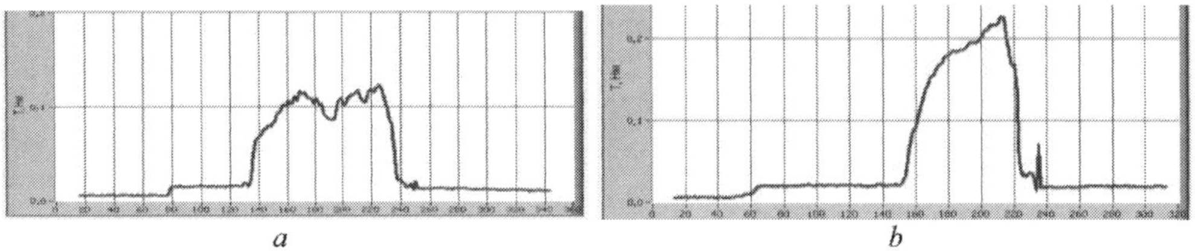

Fig. 2. The tests of the process of aramid thread SHMM-DZh axial gap kinetics: control sample (a); sample after electrophysical treatment (b).

Control thread UHMWP thread has bundle of constituent fibers after breaking. The fiber processed by electrophysical method has no significant bundles. 0.05 mm thick threads after processing have increased their tensile strength in 1.6-1.7 times, the threads 0.03 mm thick - nearly in 1.6 times. The threads elongation in 15-20 mm was noted. Decreasing the control threads thickness from 0.05 mm to 0.03 mm reduces the tear strength in 55%, the processed thread - in 65%.

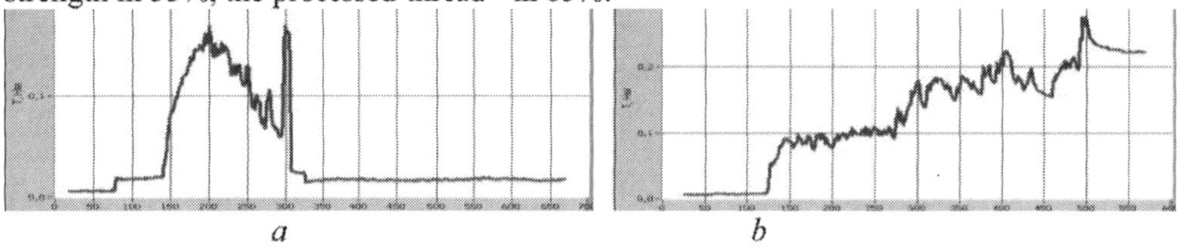

Fig. 3. The kinetics of the process of the axial gap filament UHMWP thickness of 0.05 mm control (a) and processed (b).

Based on the analysis of the results the following mechanism of improving the strength aramid fibers and tissues after exposure to electromagnetic fields can be offered. Studies [3, 4] has found that the structural units of the aramid fibers are rigid macromolecules or their aggregates - fibrils. High strength and stiffness of such fibers under tension are caused by a high degree of orientation of the macromolecules along the fiber axis and a high chemical bond dissociation energy. Stress state difference of neighboring structural elements causes an appearance of shear stresses in a boundary area between the fibers, which form the thread. As a result of uniaxial fiber's tensile is accompanied with germination of inter fibrillar cracks along the direction of the tensile force. However, the split fiber up to a certain limit is still in a state to percept the load, that is why Figure 3 (a) and (b) has a smooth character of increase to a certain meaning. Cleavage of threads apparently is accompanied by simultaneous breakage of individual the most stressed fiber fragments, after which the load is perceived by beam of fibrils and so on until the remaining fibers are unable to resist tensile load. A catastrophic complete destruction happens, accompanied by intensive splitting and fluffing of a sample. The impact of electromagnetic fields is likely to contribute to the growth of the number of links on inter - fibrillar macromolecular level, the organization of fibrils in the form of spirals, running along the fiber. Such spirals of the adjacent fibers may additionally communicate with each other at a macro level in the type «nut-screw» when a coil of a spiral enters the adjacent recess. This can happen at the initial time of loading, when the fibers are stretched and straightened and the gaps between the turns increase. With a further increase in load conditions of joint, rather than separate fiber work are formed. The occurrence of cracks that weakens the structure, mentioned above, that weakens the structure, is leveled in this case by the described action of the electromagnetic field generated in the spiral interfiber structures and additional fibrils bonds.

4 Conclusions

Short-term influence of electromagnetic field on the threads and fabric made of the materials used in the manufacture of flexible personal protective equipment such as UMWP and TSVM contributes both to increase tensile strength of the individual threads, and an increase in strength of the fabric when it is punctured. This reduces the dependence of the material on external factors, such as humidity. The reason for these results is the formation of additional zones tissue connections at micro level (fibrils and the fiber) and nanoscale (macromolecule) due to the interaction of the electromagnetic field with the macromolecules that form polymer tissue fiber , and of the fibers increase in elasticity. This improves the total tensile strength of threads and also increases the time period before the gap. Application processing in the electromagnetic field of the final elements made of aramid and UHMWP fabrics will improve the reliability and durability of aircraft crews personal protective equipment (PPE) for various purposes, as well as lung protective reinforced composite structures.

References

1. E. Kablov, Aviation materials and technology **1**, 3 (2015)
2. E. Kablov, Herald of the Russian Academy of Sciences **82**, 520 (2012)
3. Y. Shustov, A. Kurdenkova, Scientific and methodical electronic journal "Concept" **3**, 36 (2016)
4. P. Shuldeshova, G. Zhelezina, *Proceedings VIAM: electron. scientific-tech. Zh.*, 9 (2014)
5. P. Safonov, *Razrabotka optimal'nyh tehnologicheskih parametrov izgotovlenija aramidnyh tkanej tehnicheskogo naznachenija// avtoref. diss. kand. tehn. nauk* (2013)
6. E. Chistyakov, Tehnika i vooruzhenie **8** , 1(2013)
7. A. Ignatova, A. Artemov, Basic Research **6-1**, 101 (2012)
8. A. Bhatnagar, *Lightweight ballistic composites* (Woodhead Pub. Lim., Cambridge, 2006)
9. I. Zlobina, N. Bekrenev, Materials Science Forum **870**, 101 (2016)
10. I. Zlobina, N. Bekrenev, High technologies **17-2**, 25 (2016)

AUTHOR INDEX

AUTHOR INDEX